CAMBRIDGE LIBRARY COLLECTION

Books of enduring scholarly value

Zoology

Until the nineteenth century, the investigation of natural phenomena, plants and animals was considered either the preserve of elite scholars or a pastime for the leisured upper classes. As increasing academic rigour and systematisation was brought to the study of 'natural history', its subdisciplines were adopted into university curricula, and learned societies (such as the London Zoological Society, founded in 1826) were established to support research in these areas. These developments are reflected in the books reissued in this series, which describe the anatomy and characteristics of animals ranging from invertebrates to polar bears, fish to birds, in habitats from Arctic North America to the tropical forests of Malaysia. By the middle of the nineteenth century, this work and developments in research on fossils had resulted in the formulation of the theory of evolution.

The Dodo and Its Kindred

Well versed in natural history, particularly geology and ornithology, Hugh Edwin Strickland (1811–53) became fascinated by the dodo and mankind's influence on its extinction. Seeking to investigate this flightless bird and other extinct species from islands in the Indian Ocean, he invited the comparative anatomist Alexander Gordon Melville (1819–1901) to help him separate myth from reality. Divided into two sections, this 1848 monograph begins with Strickland's evaluation of the evidence, including historical reports as well as paintings and sketches, many of which are reproduced. Melville then analyses the osteology of the dodo and Rodrigues solitaire, describing his findings from dissections of the few available specimens and making comparisons with similar species. A seminal work, it correctly concluded that the dodo was more closely related to pigeons than vultures, and the book also inspired others to take up the search for new fossil evidence.

The Dodo
and Its Kindred

Or The History, Affinities, and Osteology of the Dodo, Solitaire, and Other Extinct Birds of the Islands Mauritius, Rodriguez, and Bourbon

Hugh Edwin Strickland
Alexander Gordon Melville

CAMBRIDGE
UNIVERSITY PRESS

CAMBRIDGE
UNIVERSITY PRESS

University Printing House, Cambridge, CB2 8BS, United Kingdom

Cambridge University Press is part of the University of Cambridge.
It furthers the University's mission by disseminating knowledge in the pursuit of education, learning and research at the highest international levels of excellence.

www.cambridge.org
Information on this title: www.cambridge.org/9781108078313

This edition first published 1848
This digitally printed version 2015

ISBN 978-1-108-07831-3 Paperback

The original edition of this book contains a number of colour plates, which have been reproduced in black and white. Colour versions of these images can be found online at www.cambridge.org/9781108078313

THE DODO AND ITS KINDRED.

Fac-simile of Savery's picture of the DODO, in the Royal Gallery at Berlin.

THE

DODO AND ITS KINDRED;

OR THE

HISTORY, AFFINITIES, AND OSTEOLOGY

OF THE

DODO, SOLITAIRE,

AND

OTHER EXTINCT BIRDS

OF THE ISLANDS MAURITIUS, RODRIGUEZ, AND BOURBON.

BY

H. E. STRICKLAND, M.A., F.G.S., F.R.G.S.,
PRESIDENT OF THE ASHMOLEAN SOCIETY, &c.,

AND

A. G. MELVILLE, M.D. EDIN., M.R.C.S.

"PES et CAPUT uni
Reddentur formæ."—*Hor.*

LONDON:
REEVE, BENHAM, AND REEVE, 8, KING WILLIAM STREET, STRAND.

1848.

TO

P. B. DUNCAN, ESQ., M.A.,

KEEPER OF THE ASHMOLEAN MUSEUM,

This Work is Inscribed,

AS A SLIGHT TOKEN OF REGARD AND ESTEEM,

BY HIS SINCERE FRIENDS,

THE AUTHORS.

List of Subscribers.

H. R. H. PRINCE ALBERT, K.G.
HER GRACE THE DUCHESS OF BUCCLEUCH *(two copies)*.
THE MOST NOBLE THE MARQUIS OF NORTHAMPTON, Pres. Brit. Assoc., Pres. R.S.
THE RIGHT HON. THE EARL OF DERBY, Pres. Z.S., F.R.S.

The Radcliffe Library, Oxford.
The Belfast Library.
The Edinburgh College Library.
The Signet Library, Edinburgh.
Zoological Society of London.
York Philosophical Society.
Worcestershire Nat. Hist. Society.
King's College Library, Aberdeen.
Royal Society of Arts and Sciences, Mauritius.
The Chevalier Dubus, Brussels.
The Baron de Selys Longchamps, Liege.
Admiral Mitford, Hunmanby, Yorkshire.
Sir Robert Harry Inglis, Bart., M.P.
Sir John G. Dalyell, Bart.
Sir W. R. Boughton, Bart., *(two copies)*.
Sir W. C. Trevelyan, Bart., F.R.S.
Sir William Jardine, Bart., F.R.S.E.
Sir T. Tancred, Bart.
Sir James S. Menteath, Bart.
The Very Reverend the Dean of Westminster.
Reverend The Master of University College, Oxford.
Rev. Dr. Dunbar, Applegarth, Dumfriesshire.
Rev. Professor Walker, F.R.S., Oxford.
Rev. Professor Hussey, Oxford.
Professor Daubeny, F.R.S.
Professor Bell, F.R.S.
Professor Lizars, Aberdeen.
Professor Ansted, F.R.S.
Professor J. Phillips, F.R.S.
Professor J. F. Johnston, Durham.
Professor H. Lichtenstein, Berlin.
Professor Schinz, Zurich.
Professor Goodsir, Edinburgh.
Professor Carl J. Sundevall, Stockholm.
Rev. J. Hannah, Rector of the Edinburgh Academy.
Rev. A. D. Stacpoole, New College, Oxford.
Rev. F. O. Morris, Nafferton, Yorkshire.
Rev. A. Matthews, Weston, Oxfordshire.
Rev. W. C. Fowle, Ewias Harold, Herefordshire.
Rev. W. W. Cooper, Claines, Worcester.
Rev. J. M. Prower, Pyrton, Gloucestershire.
Rev. J. Griffiths, Wadham College, Oxford.
Rev. T. Ewing, Hobart Town.
Rev. T. A. Strickland, Bredon, Gloucestershire.
Rev. W. H. Stokes, Caius College, Cambridge.
Rev. W. Little, Kirkpatrick Juxta, Dumfriesshire.
W. J. Hamilton, Esq., M.P.
R. Parnell, M.D., Edinburgh.
G. Lloyd, M.D., Warwick.
H. W. Acland, M.D., Reader in Anatomy, Oxford.
Dr. Charlton, Newcastle.
J. Scouler, M.D., Dublin.
Dr. Cogswell.
W. A. Greenhill, M.D., Oxford.
C. Hastings, M.D., Worcester.
Dr. G. Hartlaub, Bremen.
Dr. Davis, Bath.
Dr. Bennet, Sydney.
T. Horsfield, M.D., F.R.S.
Hugh Falconer, M.D., F.R.S.
Mrs. Dixon, Govan Hill, Glasgow.
Mrs. A. Smith, Edinburgh.
Mrs. C. Clarke, Matlock.
Mrs. Hodder, Leith Links, Edinburgh.
Miss Christie, Balmuto, Edinburgh.
Miss Wedderburn, Berkhill, Edinburgh.
Miss Porter, Birlingham, Worcestershire.
Miss L. Strickland, Dawlish, Devonshire.
P. B. Duncan, Esq., M.A., New Coll. Oxford. *(two copies)*.
James Yates, Esq., M.A., F.R.S.
C. Stokes, Esq., F.R.S.
John Edward Gray, Esq., F.R.S.
John Gould, Esq., F.R.S.
William Spence, Esq., F.R.S.
J. S. Bowerbank, Esq., F.R.S.
John Arrowsmith, Esq., F.R.G.S.
William Yarrell, Esq., F.L.S.
P. J. Selby, Esq. F.L.S.
Adam White, Esq., F.L.S.
G. R. Gray, Esq , F.L.S.
T. C. Eyton, Esq., F.L.S.
Major P. T. Cautley.
Lieut. John Croker.
T. B. L. Baker, Esq., Hardwick Court, Gloucester.
J. Wolley, Esq., Edinburgh.
A. Carruthers, Esq., Warmanbie, Dumfriesshire.
Andrew Murray, Esq., W.S., Edinburgh.
John M. Fenwick, Esq., Gallow Hill, Morpeth.
G. R. Waterhouse, Esq., British Museum.
W. Thompson, Esq., Belfast.
A. Johnstone, Esq., Halleaths, Dumfriesshire.
G. Shuttleworth, Esq.
W. Bell Macdonald, Esq., Rammerscales, Dumfriesshire.
Archibald Hepburn, Esq.
D. W. Mitchell, Esq., Sec. Z.S.
Robert Heddel, Esq.
H. N. Turner, Esq.
T. Stevenson, Esq., C.E., Edinburgh.
Samuel Maunder, Esq.
W. H. Lizars, Esq., Edinburgh.
J. W. Salter, Esq.
C. Winn, Esq., Nostall Priory, Yorkshire.
J. D. Murray, Esq., Murraythwaite, Dumfriesshire.
H. B. W. Milner, Esq., All Souls' College, Oxford.
W. V. Guise, Esq., Elmore Court, Gloucester.
And. Jardine, Esq., Lanrig Castle, Stirlingshire.
Edward Wilson, Esq., Lydstip House, Tenby.
John Henry Gurney, Esq.
P. L. Sclater, Esq., C.C.C., Oxford.
J. H. Wilson, Esq., Wadham College, Oxford.
Henry Deane, Esq.
George Peevor, Esq.
T. A. Knipe, Esq., Clapham.
M. Fairmaire, Paris.
C. W. Orde, Esq., Nunnykirk, Morpeth.
Samuel E. Cottam, Esq., Brazennose Street, Manchester.
H. Hussey, Esq., 6, Upper Grosvenor Street, London.
Lovell Reeve, Esq., F.L.S.
E. Benham, Esq.
F. Reeve, Esq.

PART I.

HISTORY AND EXTERNAL CHARACTERS

OF THE

DODO, SOLITAIRE,

AND OTHER

EXTINCT BREVIPENNATE BIRDS

OF

MAURITIUS, RODRIGUEZ, AND BOURBON.

BY

H. E. STRICKLAND, M.A., F.G.S.

INTRODUCTION.

Among the many remarkable results connected with Organic Life which modern Science has elicited, the chronological succession of distinct races of beings is one of the most interesting. Geology exhibits to us the vast diversity of organized forms which have supplanted one another throughout the world's history, and in dealing with this remarkable fact, we are led to search out the causes for these exits and entrances of successive actors on the stage of Nature. It appears, indeed, highly probable that Death is a law of Nature in the Species as well as in the Individual; but this internal tendency to extinction is in both cases liable to be anticipated by violent or accidental causes. Numerous external agents have affected the distribution of organic life at various periods, and one of these has operated exclusively during the existing epoch, viz. the agency of Man, an influence peculiar in its effects, and which is made known to us by testimony as well as by inference. The object of the present treatise is to exhibit some remarkable examples of the extinction of several ornithic species, constituting an entire sub-family, through Human agency, and under circumstances of peculiar interest.

The geographical distribution of organic groups in space is a no less interesting result of science than their geological succession in time. We find a special relation to exist between the structures of organized bodies and the districts of the earth's surface which they inhabit. Certain groups of animals or vegetables, often very extensive, and containing a multitude of genera or of species, are found to be confined to certain continents and their circumjacent islands.[1] In the present state of science we must be content to admit the existence of this law, without being able to enunciate its preamble. It does *not* imply

[1] To cite one instance among a thousand: the group of Humming Birds, containing hundreds of species, is exclusively confined to the American continent and the West Indian Archipelago.

that organic distribution depends on soil and climate; for we often find a perfect identity of these conditions in opposite hemispheres and in remote continents, whose faunæ and floræ are almost wholly diverse. It does *not* imply that allied but distinct organisms have been educed by generation or spontaneous development from the same original stock; for (to pass over other objections) we find detached volcanic islets which have been ejected from beneath the ocean, (such as the Galapagos for instance,) inhabited by terrestrial forms allied to those of the nearest continent, though hundreds of miles distant, and evidently never connected with them. But this fact *may* indicate that the Creator in forming new organisms to discharge the functions required from time to time by the ever vacillating balance of Nature, has thought fit to preserve the regularity of the System by modifying the types of structure already established in the adjacent localities, rather than to proceed *per saltum* by introducing forms of more foreign aspect. We need not, however, pursue this enquiry further into obscurity, but will merely refer to the law of geographical distribution, as bearing on the subject before us.

In the Indian Ocean, to the east of Madagascar, are three small volcanic islands, which, though somewhat scattered, are nearer to each other than to any neighbouring land. This circumstance gives them a claim to be regarded as a geographical group, a meagre fragment of an archipelago, although in a general sense they are connected with Madagascar, and more remotely with the African continent. In conformity with the above-mentioned relation between geographical distribution and organic structure, we find that a small portion of the indigenous animals and plants of those islands are either allied or identical with the products of Africa, a larger portion with those of Madagascar, while certain species are peculiar to the islands themselves. And as these three islands form a detached cluster, as compared to other lands, so do we find in them a peculiar group of birds, specifically different in each island, yet allied together in their general characters, and remarkably isolated from any known forms in other parts of the world. These birds were of large size and grotesque proportions, the wings too short and feeble for flight, the plumage loose and decomposed, and the general aspect suggestive of gigantic immaturity. The history of these birds was as remarkable as their organization. About two centuries ago their native isles were first colonized by Man, by whom these strange creatures were speedily exterminated. So rapid and so complete was their extinction that the vague descriptions given of them by early navigators were long regarded as fabulous or exaggerated, and these birds, almost contemporaries of our great-grandfathers, became associated in the minds of many persons with the Griffin and the Phœnix of mythological antiquity. The aim of the present work is to vindicate the honesty of the rude voyagers of the 17th century, to collect together the scattered evidences which we possess, to describe and depict the few anatomical fragments of these lost species which are still extant, to incite the scientific traveller to search for further evidences, and to infer from the data before us the probable rank of these birds in the System of Nature.

These singular birds, which for distinction we shall henceforth designate by the technical name *Didinæ,* furnish the first clearly attested instances of the extinction of organic species through human agency. It has been proved, however, that other examples of the kind have occurred both before and since;[1] and many species of animals and of plants are now undergoing this inevitable process of destruction before the ever-advancing tide of human population.[2] We cannot see without regret the extinction of the last individual of any race of organic beings, whose progenitors colonized the pre-adamite Earth; but our consolation must be found in the reflection, that Man is destined by his Creator to "be fruitful and multiply and replenish the Earth and subdue it." The progress of Man in civilization, no less than his numerical increase, continually extends the geographical domain of Art by trenching on the territories of Nature, and hence the Zoologist or Botanist of future ages will have a much narrower field for his researches than that which we enjoy at present. It is, therefore, the duty of the naturalist to preserve to the stores of Science the knowledge of these extinct or expiring organisms, when he is unable to preserve their lives; so that our acquaintance with the marvels of Animal and Vegetable existence may suffer no detriment by the losses which the organic creation seems destined to sustain.

In the case of the *Didinæ,* it is unfortunately no easy matter to collect satisfactory information as to their structure, habits, and affinities. We possess only the rude

[1] As instances, I may mention the *Cervus megaceros,* or Irish Elk, and the *Bos primigenius,* or Urus, destroyed in ancient, and the *Rytina Stelleri,* or Northern Dugong, in modern times.

[2] Among animals whose doom is probably not far distant are the *Bison priscus,* or Aurochs, (preserved only by imperial intervention in the Bialowicksa forest, whence the Czar has lately enriched the London Zoological Gardens with a living pair); the *Nestor productus,* (a Parrot originally from Phillip's Island near Norfolk Island, where it is now destroyed, though a few individuals, which refuse to propagate, still survive in cages); the two (not improbably *three*) species of *Apteryx*; and the almost equally anomalous burrowing Parrot, *Strigops habroptilus,* of New Zealand; &c.

descriptions of unscientific voyagers, three or four oil paintings, and a few scattered osseous fragments, which have survived the neglect of two hundred years. The paleontologist has, in many cases, far better data for determining the zoological characters of a species which perished myriads of years ago, than those presented by a group of birds, several species of which were living in the reign of Charles the First.

We shall find it convenient to treat of each island, and of its ornithic productions, separately. And, first, of the best known and most celebrated of these creatures, the brevipennate bird of Mauritius, the Dodo.

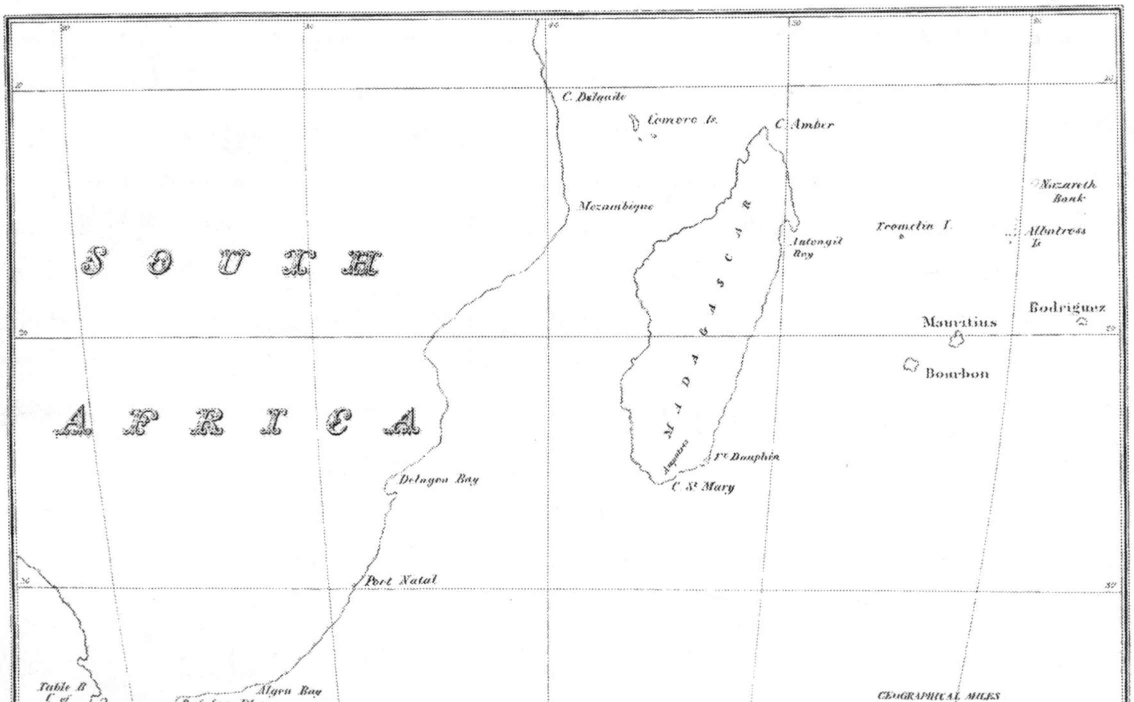

CONTENTS.

THE

NATURAL HISTORY

OF THE

DODO, SOLITAIRE, &c.

PART I.

CHAPTER I.

The Brevipennate Bird of Mauritius, the Dodo, (*Didus ineptus* of Linnæus.)

Section I.—*Division of the subject—Historical evidences—Discovery of the Islands—Voyage of Van Neck; of Heemskerk and Willem—Dodo's leg at Leyden—Voyage of Matelief; of Van der Hagen; of Verhuffen; of Van den Broecke; of Herbert; of Cauche—Dodo exhibited in London—Account given by Tradescant; by Piso; by Hubert; by Olearius—Harry's Voyage—Extinction of the Dodo—Negative character of modern evidence.*

Most persons are acquainted with the general facts connected with that extraordinary production of Nature, known by the name of the *Dodo,*—that strange abnormal Bird, whose grotesque appearance, and the failure of every effort made for the last century and a half to discover living specimens, long caused its very existence to be doubted by scientific naturalists. We possess, however, unquestionable evidence that such a bird formerly existed in the small Island of Mauritius, and it is ascertained with no less certainty that the species has been utterly exterminated for a period of nearly two centuries.

The evidences which we possess respecting the Dodo, may be conveniently arranged on the plan adopted by Mr. Broderip, in his valuable essay on the subject,[1] by dividing them into *historical, pictorial,* and *real.*

[1] Penny Cyclopædia vol. ix. p. 47.

In enumerating the HISTORICAL EVIDENCES on this subject, I shall confine myself to such authorities as appear to be original and independent of each other. The facts recorded by these witnesses have been transcribed and often confounded by a multitude of compilers, and it is therefore indispensable to our purpose to attend mainly to the statements of original observers, and to refer only incidentally to the remarks of commentators. It has also appeared desirable not merely to translate, but to reprint the exact words of those brave old voyagers, who in the infancy of nautical and medical science, encountered a vast amount of peril and suffering, and yet found means to observe and record the natural wonders which came in their way.

Compilers are unanimous in stating that the Islands of Mauritius and Bourbon were first discovered by Mascaregnas, a Portuguese, who gave his own name to the latter island, and called the former Cerne.[1] I have not been able to find the original authority for this statement, though it is probably founded on fact. Castagneda, Osorio, Barros, Roman, Lafitau, and the other authors who treat of the Portuguese conquests in India, record the exploits of Pedro Mascaregnas, and of two or three other persons of the name, but apparently make no allusion to the discovery of these islands, which, indeed, lay completely out of the ordinary track of the Portuguese navigators. There is also a great discrepancy in the date assigned to the discovery, which one writer[2] fixes at 1502; a second,[3] at 1505; a third,[4] at 1542; and a fourth,[5] at 1545.[6] Be this as it may, it seems clear that nothing definite is recorded of Mauritius or its productions until 1598, when the Dutch under Jacob Cornelius Neck, or Van Neck, finding it uninhabited, took possession, and changed its name from Cerne to Mauritius.

[1] The Portuguese discoverers appear to have named this island *Cerne*, from an utterly untenable notion that it might be the *Cerne* of Pliny (Hist. Nat. vi. 36, and x. 9.), an island which, according to the usual punctuation of the text, lay off the Persian Gulf, but was more probably on the West Coast of Africa (see A. de Grandsagne's edition of Pliny, Paris, 1829, vol. iv. p. 143, and vol. v. p. 344). Later authors, however, from Clusius downwards, insist that the Portuguese called it *Cerne* or *Cisne*, i. e. *Swan Island*, from the Dodos, which they compared to Swans (see Clusius, Exotica, p. 101). The statement that Vasco de Gama, in 1497, discovered, sixty leagues beyond the Cape of Good Hope, a bay called after San Blaz, near an island full of birds with wings like bats, which the sailors called *Solitaries* (De Blainville, Nouv. Ann. Mus. H.N., and Penny Cyclop. DODO, p. 47.) is wholly irrelevant. The birds are evidently Penguins, and their wings were compared to those of bats, from being without developed feathers. De Gama never went near Mauritius, but hugged the African Coast as far as Melinda, and then crossed to India, returning by the same route. This small island inhabited by Penguins, near the Cape of Good Hope, has been gratuitously confounded with Mauritius. Dr. Hamel, in a Memoir in the *Bulletin de la Classe Physico-mathématique de l'Acad. de St. Pétersbourg*, vol. iv. p. 53, has devoted an unnecessary amount of erudition to the refutation of this obvious mistake. He shews that the name *Solitaires*, as applied to Penguins by De Gama's companions, is corrupted from *Sotilicairos*, which appears to be a Hottentot word.

[2] Ersch and Gruber's Encyclopädie. [3] Grant's Mauritius. [4] Penny Cyclopædia.

[5] Du Quesne in Leguat's Voyage, on the authority of a stone pillar, placed in Bourbon by the Portuguese.

[6] In one of De Bry's maps, which illustrates the *first* Dutch expedition of 1595–1597, these islands are indicated as "I. de Mascarenhas."

Comment nous avons (*sur l'Isle* Maurice, *autrement nommée* do Cerne) *tenu mesnage*. *No*. 2.

Fac-simile of Plate 2 of Van Neck's Voyage.

1. In the published narrative of this Voyage,[1] it is stated that they found in the island a variety of pigeons, parroquets, and other birds, among which were some which they denominated *Walckvögel*, the size of swans, with a large head furnished with a kind of hood; no wings, but in place of them three or four small black quills; and the tail consisted of four or five curled plumes of a grey colour. The Dutch sailors called them *Walckvögel*, or *disgusting birds*, from the toughness of their flesh, as might be expected in the strongly developed crural muscles of a cursorial bird, though they found the pectoral muscles more palatable. The ample supply of turtle-doves also caused the *Walckvögel* to be the less esteemed.

The following is De Bry's version of this account, and in cases where the French translation (Amsterdam, 1601) differs in sense, the latter is quoted also:

"Insula dicta præterquam quod terræ nascentibus feracissima sit, volucres etiam copiosissimas alit, ut sunt, turtures, qui tanta ibi copia obversantur, ut terni nostrum dimidii diei spatio 150 aliquando ceperimus, plures facilè prehensuri manibus, aut cœsuri fustibus, si illorum onere non nimium nos pressos sensissemus. Cærulei quoque psittaci ("*parroquets gris*," Fr.) ibi frequentes sunt ut et aves aliæ: præter quas genus aliud quoque grandius conspicitur, cygnis nostris majus ("*de la grandeur de nos Cignes*," Fr.) capitibus vastis, et pelle ex dimidia parte q. cucullis investitis. Hæ aves alis carent: quarum loco tres quatuorve pennæ nigriores prodeunt. Caudam constituunt pauculæ incurvæ pennæ teneriusculæ, ("*au lieu du Cap, ont ils quatre ou cincq plumettes crespues*," Fr.) colorem cineris referentes. Has nos *Walckvögel* appellitabamus, hanc ob causam, quod quo longius seu diutius elixarentur, plus lentescerent et esui ineptiores fierent. Illarum tamen ventres et pectora saporis jucundi et masticationis facilis erant; ("*voire fort coriaçes, mais estoient medicine pour l' estomach et la poictrine*," Fr.). Appellationis causa altera erat, quod turtures ibi optabili copia nobis sufficerent, saporis longe gratioris et suavioris."—De Bry, pars V. p. 7.

The quaint old print, of which a fac-simile is annexed, exhibits the voyagers revelling in the abundance of this virgin isle. I will not spoil by translation the refreshing simplicity of the Batavo-Gallic description which accompanies it.

"*Declaration de ce qu' avons veu et trouvé sur l' Isle Maurice, et de ce qui est par nous executé.* No. 2.

"1. Sont Tortues qui se tiennent sur l' haut pays, frustez d'aisles pour nager, de telle grandeur, qu' ils chargent ung homme et rampent encore fort roidement; prennent aussi des Escriuisses de la grandeur d'un pied, qu' ils mengent.

[1] The earliest account of this voyage which I have seen, was published in folio at Amsterdam, by Corneille Nicolas in 1601, and a second edition in 1609, both of which are bound up in a folio volume of rare tracts, preserved in the Radcliffe Library. It is entitled 'Le second Livre, Journal ou Comptoir, contenant le vray Discours et Narration historique du voyage faict par les huict Navires d' Amsterdam au mois de Mars l' An 1598 soubs la conduitte de l' Admiral Jaques Corneille Necq, et du Vice-Admiral Wibrant de Warwicq.' Dutch and German editions were published at the same time, the latter by Hulsius, Nürnberg, 1602, and Frankfort, 1605; a Latin translation of it occupies the fifth part of De Bry's India Orientalis, 1601, and an English version appeared the same year in London. Editions were also published in quarto at Amsterdam in 1648 and 1650; M. de Blainville is therefore in error when he states (Nouv. Ann. Mus. H. N. vol. iv. p. 4) that the first account of this voyage was published at Rouen in 1725.

"2. Est ung oiseau, par nous nommé *Oiseau de Nausée,* à l' instar d' une Cigne, ont le cul rond, couvert de deux ou trois plumettes crespues, carent des aisles, mais en lieu d' icelles ont ilz trois ou quatre plumettes noires; des susdicts oiseaux avons nous prins une certaine quantité, accompagné d' aucunes Tourturelles, et autres oiseaux, qui par noz compaignons furent prins, la premiere fois qu' ils arrivoyent au pays, pour chercher la plus profonde et plus fraische Riviere, et si les navires y pourroyent estre sauvez, et retournerent d' une grande joye, distribuant chasque navire, de leur Venoison prins, dont nous partismes le lendemain vers le port, fournismes chasque navire d'un Pilote de ceux qui auparavant y avoyent esté; avons cuict cest oiseau, estoit si coriace que ne le povions asses bovillir, mais l' avons mengé a demy cru. Si tost qu' arrivames au port, envoya le Vice-Admiral nous, avecq une certaine troupe au pays, pour trouver aucun peuple, mais n'ont trouvé personne, que des Tourturelles et autres en grande abondance, lesquels nous prismes et tuames, car veu qu' il n' y eust personne qui les effraia, n' avoient ilz de nous nulle crainte, tindrent lieu, se laisserent assomer. En somme c' est un pays abondant en poisson et oiseaux, voire tellement qu' il excella tous les autres audit voyage.

"3. Un Dactier, dont les feüilles sont si grandes qu' un homme s' en peult guarantir contre la pluie sans se mouillir, et quand on y forre un trou, et le mette en broche y sort il du vin, comme vin Secq, amiable et doux: mais quand on le gard trois ou quatre jours, commenc' il a aigrer, et pourtant est il nommé vin de Palmite.

"4. Est un oiseau de nous nommé *Rabos Forcados,*[1] a cause de leur queuë en forme d' une Force, fort domptez, et quand on les extend, ont ils bien la longeur d' une brassée, a long becq, tous quasi noirs, ayants une poictrine blanche, prennent du poisson volant, qu' ils mengent, aussi les boyaux des poissons et oiseaux, comme avons experimenté a ceux qu' avions prins, car quand nous les apprestames, et dejettames les entrailles, engloutirent et devoroyent ils lesdicts entrailles et precordes de leurs confreres. Estoyent fort coriaces en cuisant.

"5. Est un oiseau de nous nommé le Corbeau Indien,[2] ayant la grandeur plus d' une fois que les Parroquets, de double et triple couleur.

"6. Un arbre sauvage, auquel nous avons mis (pour la souvenance si y pourroyent arriver aucuns navires) un aisselet, orné des armoires d' *Hollande, Zélande,* et d' *Amsterdam,* a fin qu' autres arrivants audit lieu, pourroyent veoir que les Hollandois y avoyent esté.

"7. Cecy est un Palmite. Bonne partie de ces arbres, furent par nos compagnons abatus, et en taillerent cest esclat, quotée de la lettre A, bonne remedée pour la maladie aux membres, de la longueur de deux ou trois pieds, par dedans tout blanc; douce; aucuns en mangerent bien sept ou huict.

"8. Est une Chauvesouris, testue en forme de Marmelot, volent icy en grande multitude, se pendent en grand nombre aux arbres, ont a la fois un combat entr' eux, en se mordants.

"9. Icy dressa le Mareschal une Forge, et pancha la ferraile, repara aussi certain fer qui fust es navires.

"10. Sont Cabannes par nous illecq construits d' arbres et feüilles, pour ceux qui aidoyent le Mareschal et Tonnelier a besoigner; pour partir avec la premiere commodité.

"11. En ce lieu fit nostre Ministre Philippe Pierre Delphois homme syncere et candide, une Presche fort severe, sans exception de personne, deux fois sur la ditte Isle, devant le disner y alla l' une

[1] This bird is the *Fregata aquila,* Lin.

[2] A species of *Buceros.*

partie, et apres le disner l' autre. Icy fut Laurent (*Madagascarois*) baptisé, accompagné encore d' un ou deux des nostres.

"12. Icy fismes estude de pescher, et en prismes une quantité incroyable, voire en prismes d' un seul coup bien deux et demie tonneaux, touts de diverses couleurs."

A shorter and less complete narrative of this voyage seems to have been published in German, which is translated[1] in the fourth part of De Bry's 'India Orientalis,' 1601, p. 105. in which the Walckvögel are briefly mentioned as follows:

"Eodem quoque loco aves plurimæ inveniuntur, tam grandes ut geminos cycnos æquent. Has *Walchstocken* seu *Walckuëgels* nominabant, quarum carnes esu haud incommodæ erant. Sed cum pariter ibidem magna copia Columbarum et Psittacorum appareret, quæ adiposæ et mansu suavissimæ essent, socii nostri, grandioribus fastiditis, delicatiores et teneriores aves elegerunt et ærumnas suas illarum mactatione diluerunt."

These birds are also professedly represented in plate III. of the same work, but as the figures are evidently copied from *Cassowaries,* they are of no authority, and I do not therefore reproduce them here. In the description, however, at the foot of this plate is an important statement, *if true*; viz., that the voyagers brought one of these birds with them to Holland. "In eadem insula Psittacorum Columbarumque numerum quoque maximum repererunt, tam cicurum ut fustibus eas prostraverint. Sed et aliæ ibidem aves visæ sunt, quas *Walckvögel* Batavi nominarunt, et *unam secum in Hollandiam importarunt.*" But as no contemporary author, not even the diligent Clusius, makes any further allusion to the importation of so remarkable a bird, it is possible that De Bry, or his authority, may have confounded the history, no less than the portrait, of the Cassowary with that of the Dodo, for it is well known that a live Cassowary was brought in 1597 to Holland, where it attracted much attention (Clusius, Exotica, p. 97). There are, however, as I shall afterwards show, strong grounds for believing that a living Dodo was really brought to Holland some time during the first quarter of the 17th century.

It would appear from the 'Exotica' of Clusius, 1605, that a third account of this voyage had been published in his time, which seems to be unknown to British bibliographers. Nor is this any marvel, when we consider how little Dutch literature is studied in this country, and how deficient are the best British libraries in the works of our enterprising neighbours in Holland. Clusius's figure of the Dodo is evidently distinct from, and more accurate than, the one given by Van Neck (*supra*, plate II. fig. 2.), and is copied, he says, from a published account of Van Neck's voyage. He adds that the beak was thick and

[1] Such at least is the inference from the words "omnia ex Germanico Latinitate donata," in De Bry's title page. But Camus in his 'Memoire sur la Collection des grands et petits Voyages,' Paris, 1802, p. 212. considers the account of Van Neck's Voyage in Part IV. of De Bry, to be only an abridgment of that given *in extenso* in Part V., and not a translation of a separate narrative. He also is of opinion that the first four plates of Part IV. have been composed by De Bry from the description given by the voyagers; and certainly there is a touch of the marvellous about them, which favours this idea.

long, yellowish next the head, with a black point. The upper mandible was hooked, the lower had a bluish spot in the middle between the yellow and black part, the bird was covered with thin and short feathers, the hinder part was very fat and fleshy, the legs were thick, covered to the knee with black feathers, the feet yellowish, the toes three before and one behind. He further states, that stones were found in the gizzards of these birds, and that he saw two of these stones in Holland, one of which, about an inch in length, he has figured. His original words are as follows :—

"Cap. IV. *Gallinaceus Gallus peregrinus.* Ex octo navibus illis quæ anno 1598, Aprili mense, ex Hollandiâ solvebant, &c., quinque . . . . montosam quandam insulam in conspectu habuerunt, ad quam lætabundi cursum converterunt. . . . . Dum in insulâ hærent, varii generis aves observabant; atque inter illas valdè peregrinam, cujus iconem rudi arte delineatam in Diario totam illius navigationis historiam continente, quod reduces cudi curabant, conspiciebam, ad cujus normam est expressa quam hoc capiti propono.

"Illa porro avis peregrina Cygnum quidem magnitudine æquabat aut superabat, sed ejus forma longè diversa: ejus etenim caput magnum, tectum veluti quâdam membranâ cucullum referente; rostrum præterea non planum, sed crassum et oblongum, subflavi coloris parte capiti proximâ, cujus extimus mucro niger, superior quidem ejus pars sive prona adunca et curva, in inferiore verò sive supinâ subcærulea macula mediam partem inter flavam et nigram occupabat. Raris et brevibus pennis tectam esse aiebant, et alis carere, sed earum loco quaternas aut quinas dumtaxat longiusculas nigras pennas habere: posteriorem autem corporis partem præpinguem et valdè crassam, in quâ pro caudâ quaternæ aut quinæ crispæ convolutæque pennulæ cineracei coloris: crura illi potiùs crassa esse quàm longa, quorum superna pars genu tenus nigris pennulis tecta, inferior cum pedibus subflavi coloris; pedes verò in quatuor digitos fuisse divisos, ternos longiores antrorsùm spectantes, quartum breviorem retrorsùm conversum, omnesque nigris unguibus præditos. . . . . . . Nautæ huic avi nomen indebant suo idiomate *Walgh-vogel,* hoc est, nauseam movens avis, partim quod post diuturnam elixationem, ejus caro non fieret tenerior, sed dura permaneret et difficilis concoctionis, (excepto ejus pectore et ventriculo, quæ non contemnendi saporis esse comperiebant,) partim quod multos turtures nancisci poterant, quos delicatiores et ori magis gratos reperiebant: nihil igitur mirum si præ illis hanc avem contemnerent, et eâ se facilè carere posse dicerent. In ejus porrò ventriculo quosdam lapillos inventos aiebant, quorum binos huc perlatos conspiciebam apud ornatissimum virum Christianum Porretum, eosque diversæ formæ, unum plenum et orbicularem, alterum inæqualem et angulosum, illum uncialis magnitudinis, quem juxta pedes avis exprimendum curabam, hunc majorem et graviorem, utrumque cineracei coloris; eos ab ave in maris littore lectos, deinde devoratos fuisse verisimile est, non in ejus ventriculo natos."—*Exotica*, p. 99.

2. In 1601 two fleets of Dutch ships, one commanded by Wolphart Harmansen, or Harmansz, and the other by Jacob Van Heemskerk, sailed for the East Indies, but soon separated. Harmansen's ships touched at Mauritius in their way, but in the published accounts of his voyage no mention of Dodos occurs. His companion Heemskerk, however, remained nearly three months in Mauritius, on his homeward voyage in 1602, and in a journal kept by Reyer Cornelisz, and printed in the 'Begin ende voortgang van de Vereenighde Nederlantsche Geoctroyeerde Oostindische Compagnie' (oblong 4to, 1646, s. l.) vol. i., at p. 30 of Van der Hagen's Voyage, we read of "Wallichvogels" or Dodos, among a variety of other game:—

"Op het lant onthouden haer Schiltpadden, *Wallichvogels,* Flamencos, Gansen, Eendt-vogels, Velt-hoenders, soo groot as kleyne Indiaensche Ravens, Duyven, daer onder sommighe met roo steerten, (van de welcke menig man sieck geweest is,) grauwe ende groene Papegayen, met lange steerten, waer van datter sommighe ghevangen werden."

3. One of the Captains who sailed in the fleet of Heemskerk and Harmansz, named Willem van West-Zanen, has left a journal, which apparently was not published until 1648, when it was edited and enlarged by H. Soeteboom.[1] In 1602 Willem sailed from Batavia with five richly laden ships, commanded by Admiral Schuurmans, and stayed a considerable time at Mauritius.[2] He makes repeated mention of Dod-aarsen, or Dodos, and though his account seems to have been somewhat amplified by his editor Soeteboom, yet it contains some original and important particulars. The sailors appear, on this occasion, to have revelled in Dodos, without suffering from surfeit, like Van Neck's crew. If the statements are correct that three or four, and in one instance two, of these birds furnished an ample meal for Willem's men, the bulk of the Dodo must have been prodigious, and might well have equalled fifty pounds weight, as asserted by Sir T. Herbert. As this tract is very rare, I will extract, in full, the passages which mention these birds, and annex a literal translation.

[1] This tract is entitled 'Derde voornaemste Zee-getogt (der verbondene vrye Nederlanderen) na de Oost-Indien, gedaan met de Achinsche en Moluksche Vloten, onder de Ammiralen Jacob Heemskerk en Wolfert Harmansz. In den Jare 1601, 1602, 1603. Getrocken Uyt de naarstige aanteekeningen van Wiliem van West-Zanen, Schipper op de Bruin-Vis, en met eenige noodige byvoegselen vermeerdert, door H. Soete-Boom. 4to. Amsterdam, 1648.' (Brit. Mus. $\frac{566}{5}$ f. 15.)

[2] After leaving Mauritius, Schuurmans returned to Holland in company with Harmansen and Garnier, Heemskerk's Vice-Admiral, in the spring of 1603. So that Clusius is mistaken in saying (*Exotica, p.* 101,) that this expedition was commanded by Van Neck, as the latter did not return from his second voyage until some years afterwards.

"De Vogelen (daar 't van vol is) zijn van allerhande slag: Duyven, Papegayen, Indische-Ravens, Sparwers, Valken, Lijsters, Vlen, Swaluwen, en menigten van 't kleyn gevleugelt goet; witte en swarte Reygers, Gansens, Eent-Vogels, Dod-aarsen, Schil-padden, Koeyen vander zee."—fol. 19, p. 2.

"Waren de Scheep-lieden alle dagen uyt om Vogelen en meer andere gedierten (diese op 't Landt vinden konden) te jagen, daar benevens hieldense nau op, met de Zegens, Hoeken, en andere vissing in de weer te zijn; viervoetige gedierten, uytgezondert Katten, zijnder niet, de onse hebben namaels daar Bocken, Geyten en Verkens op-geplant: De Reygeren toonden haar ongetemder als andere Vogelen, waren niet wel te krijgen, vermits haar vlugt in de dichte tacken der Boomen; zy grepen Vogelen by sommige Dod-aarsen, by sommige Dronten genaamt; kregen den naam van Wallich-Vogels, ten tijden dat Jacob van Nek hier was, om datse door t' lang zieden naulijx murruw wilden, tay en hard bleven, uytgesondert de borst en maag die seer goet waren, ook om datse door de overvloedige Tortel-duyfjes (diese konde bekomen) genoegsaamde de walg kregen van de gemelde Dod-aarsen; haar afbeeltsel is in de voorige Plaat; sy hebben groote hoofden, en daar kapkens op, zijn sonder vleugelen en staarten, hebben alleen ter zyden kleine wiekxkens, achter vier of vijf veerkens, wat meer verhieven van de andere; hebben bekken en voeten, en gemenelijk in de maag een steen eens vuysten groote hebbende."—fol. 21, p. 1.

"De Dod-aarsen met haar ronde stuyten, mosten (om datse wel gevoedt waren) mede stuyt keren; 't was al in rep en roer wat sig maar reppen kond, de Visschen die voor eenige jaren vredig leefden, wierden in de diepste water-kuylen na-gejaagt," &c.—Fol. 21, p. 2.

"Den 25 (Julius) bracht Willem met zijn matrosen eenige Dod-aarsen die seer vet waren; Scheep, al't scheepvolk, hadden aan drie of vier tot een maal-tijdt genoeg te kluyven, en daar schoot noch over. . . . . . . Sie schikten gerookte Vis, en ook gesouten Dod-aarsen, nevens Land-Schil-padden, en andere Vogelen, aan boordt, welke voor-sorg daar na wel te bate quam. Waren hier mede nog eenige dagen doende en besig aan 't Schip te brengen; de Matrosen van Willem brachten op den 4 van Oegst-maandt 50 grote Vogelen in de *Bruyn-Vis*, hier onder waren 24 of 25 Dod-aarsen, so groot en swaar datser ter maaltijd geen twee dar van opeten mogten, al watter voorts over was, wierd' in 't sout gesmeten."—Fol. 22, p. 2.

"'S anderen-daags toog Hogeveen (Willems Coopman) met vier matrosen uyt de tent, versien met stocken, netten, mosqueten, en ander gereetschap, op de Jacht, rende Heuvel en Berg op, liepen Bosch en Valey door, en vingen in de drie dagen datse uyt waren by de ander-half-hondert Vogelen, en onder de selve wel 20 Dronten of Dod-aarsen, diese alle 't Scheep brachten en in 't sout staken, sulx warense vorder, nevens 't andere volk vande vloot, in 't Vogelen en Visschen besig."—Fol. 23, p. 1.

TRANSLATION.

"The birds (of which the island is full) are of all kinds: Doves, Parrots, Indian Crows, Sparrows, Hawks, Thrushes, Owls (?), Swallows, and many small birds; white and black Herons, Geese, Ducks, *Dodos*, Tortoises, Sea-cows.

"The sailors were out every day to hunt for birds and other game, such as they could find on the land, while they became less active with their nets, hooks, and other fishing tackle. No quadrupeds occur there except Cats, though our countrymen have subsequently introduced Goats and Swine. The Herons were less tame than the other birds, and were difficult to procure, owing to their flying amongst the thick branches of the trees. They also caught birds which some name *Dod-aarsen*, others *Dronten;* when Jacob van Neck was here, these birds were called *Wallich-Vogels*, because even a long boiling would scarcely make them tender, but they remained tough and hard, with the exception of the breast and belly, which were very good; and also, because, from the abundance of Turtle-doves which the men procured, they became disgusted with the Dodos. The figure of these birds is given in the accompanying plate; they have great heads, with hoods thereon; they are without wings or tail, and have only little winglets on their sides, and four or five feathers behind, more elevated than the rest. They have beaks and feet, and commonly in the stomach a stone the size of a fist.[1] . . . . .

"The Dodos, with their round sterns, (for they were well fattened,) were also obliged to turn tail; everything that could move was in a bustle; the fish, which had lived in peace for many a year, were pursued into the deepest water-pools. . . . . .

"On the 25th July, Willem and his sailors brought some Dodos which were very fat; the whole crew made an ample meal from three or four of them, and a portion remained over. . . . . . . They sent on board smoked fish, salted Dodos, Land-tortoises, and other game, which supply was very acceptable. They were busy for some days bringing provisions to the ship. On the 4th of August Willem's men brought 50 large birds on board the *Bruyn-Vis;* among them were 24 or 25 Dodos, so large and heavy, that they could not eat any two of them for dinner, and all that remained over was salted.

"Another day, Hogeveen (Willem's supercargo) set out from the tent with four seamen, provided with sticks, nets, muskets, and other necessaries for hunting. They climbed up mountain and hill, roamed through forest and valley, and during the three days that they were out they captured another half hundred of birds, including a matter of 20 Dodos, all which they brought on board and salted. Thus were they, and the other crews in the fleet, occupied in fowling and fishing."

This account is accompanied by a very rude plate, intended to represent the "Scheep-lieden" killing Dodos; but as the artist has evidently taken Penguins as his models, I do not repeat this engraving. At the foot of the plate are these lines:—

"Victali soektmen hier en vlees van't pluim gediert,
Der pallembomen sap, de dronten rond van stuiten,
't Wylmen de papegai hout dat hij piept en tiert,
En doet dat and're meer ook raaken inder miuten."

Which may be thus Englished:—

"For food the seamen hunt the flesh of feathered fowl,
They tap the Palms, the round-sterned Dodos they destroy,
The Parrot's life they spare that he may scream and howl,
And thus his fellows to imprisonment decoy."

It is not easy to determine the date when the synonymous words *Dodars,* from which our name Dodo is derived, and *Dronte* were first introduced. The earliest apparent authority for their use is this voyage of Willem van West-Zanen, but his Journal, though written in 1603, seems to have been unpublished till 1648, and these names may therefore have been interpolated among the other alterations made in Willem's text by his editor Soeteboom. Matelief's Journal, again, which speaks of *Dodaersen,* otherwise *Dronten,* was written in 1606, and Van der Hagen's in 1607, but I have seen no edition of either work earlier than

[1] This description is evidently extracted from Matelief's Voyage.—Vide infra, p. 17.

1646, and these words may therefore be likewise due to the officiousness of editors. The earliest use of the word *Dodars* may, after all, date from 1613, when Verhuffen's Voyage was published; here, however, it occurs under the corrupt form of *Totersten*. There is little doubt that the name is derived from *Dodoor*, which in the Dutch language means a *sluggard*, and is very applicable to the lazy habits and appearance of this bird. *Dodaers* is not improbably a cant word among Dutch sailors, analogous to our term "*lubber*," and perhaps aims at expressiveness rather than elegance. Sir Thomas Herbert was the first to use this name in its modern form of *Dodo*. He tells us that it is a Portuguese word; and, in fact, we find that *doudo* in the last-named language, means "*foolish*" or "*simple*." But as none of the Portuguese voyagers appear to have mentioned the Dodo, nor even to have visited Mauritius subsequently to their first discovery of the island, such a derivation is highly improbable. It seems far more likely that *Dodars* is a genuine Dutch word, and that the pedantic Sir Thomas, who delighted in far-fetched etymologies, altered it to *Dodo* in order to make it fit with his philological theories.

The derivation of the word *Dronte*, is still more obscure than that of *Dodo*. German, Dutch, and Scandinavian dictionaries are alike unconscious of such a word. Can it be synonymous in meaning with *Dodoor*, and allied to the English *drone*, in German, *drohne*?

4. In 1605, Clusius saw in the house of Pauwius, a professor at Leyden, a Dodo's leg, which he describes as having the tarsus a little more than four inches long, and nearly four inches in circumference, covered with thick yellowish scales, broad in front, and smaller and darker coloured behind. The middle toe to the nail, was a little over two inches long, the two next were under two inches, and the hind toe one inch and a half; all the claws were thick, black, and less than an inch long, except that of the back toe, which exceeded an inch. All trace of this specimen is now lost. It is not mentioned in the 'Catalogue of all the cheifest rarities in the publick Theater and Anatomie-Hall of the University of Leiden,' 4to., Leiden, 1678; nor in a later edition of that Catalogue, published by Gerrard Blancken, in 1707; nor in the apparently contemporary tract entitled 'Res curiosæ et exoticæ in Ambulacro Horti Academici, Lugduno-Batavi conspicuæ;' nor in two old catalogues of wet preparations preserved at Leyden, all which are bound together in a volume in the Bodleian Library (Linc. F. 1. 31.); and M. de Blainville tells us that he sought for it in the Museums of Leyden and Amsterdam without success. The following is Clusius' account:—

"Verumenimverò, concinnatâ et descriptâ jam quâ potui fide hujus avis historiâ, illius crus genu tenus rescissum apud Cl. V. Petrum Pawium, primarium artis medicæ in Academiâ Lugduno-Batavâ Professorem videre contigit recens è Mauritii Insulâ relatum. Erat autem non valdè longum, sed à genu usque ad pedis inflexionem paullò plus quàm quatuor uncias superabat; ejus verò crassitudo magna, ut cujus ambitus pænè quatuor uncias æquabat, crebrisque corticibus seu squamis tectum erat, pronâ quidem parte latioribus et flavescentibus, supinâ verò minoribus, et fuscis: pedis etiam digitorum prona pars singularibus iisque latis squamis prædita, supina autem tota callosa: digiti satis breves pro tam crasso crure; nam maximi sive medii ad unguem usque longitudo binas uncias non admodùm superabat, aliorum duorum illi proximorum vix binas uncias æquabat, posterioris sescunciam: omnium verò ungues crassi, duri, nigri, minùs unciâ longi, sed posterioris digiti longior reliquis, et unciam superans."—*Exotica,* lib. v. cap. iv. p. 100.

5. Cornelius Matelief, a Dutch Admiral, arrived at Mauritius in 1606, and after alluding in his Journal to the abundance of birds in the island, he proceeds:—

"On y trouve encore un certain oisean, que quelques-uns nomment Dodarse, on Dodaersen: d'autres lui donnent le nom de Dronte. Les premiers qui vinrent en cette isle les nommèrent Oiseaux de dégoût, parce qu'ils en pouvoient prendre assez d'autres, qui étoient meilleurs. Ils sont aussi grands qu'un cigne, et couverts de petites plumes grises, sans avoir d'ailes ni de queuës, mais seulement des ailerons aux côtés, et 4 ou 5 petites plumes au derrière, un peu plus élevées que les autres. Leurs piés sont grands et épais, leur bec et leurs yeux fortlaids, et ordinairement ils ont dans l'estomac une pierre aussi grosse que le poing."—*Recueil des Voiages de la Comp. des Ind. Or.* vol. iii. p. 214.

The Dutch version of this account is as follows:—

"Men vinter ooc sekeren vogel, die van sommige *Dodaersen* genaemt wort, van andere *Dronten,* de eerste die hier arriveerden hietense Walgh-voghels, om datse andere genoech konden krijgen. Dese zijn so groot als een Swane, met kleyne grauwe veerkens, sonder vleugelen oft staert, hebben alleen ter zijde kleyne wiecken, ende achter vier of vijf veerkens, wat meer verheven als de andere, hebben groote dicke voeten, met een grooten leelijcken beck en oogen, ende hebben gemeenlijck inde mage een steen so groot as een vuyst. Sy zijn redelijck om te eten, maer t' beste datter aen is, is de maeg."—Begin ende voortgangh der Vereenighde Nederl. Geoctroyeerde Oostindische Compagnie, vol. ii., Matelief's Voy. p. 5.

6. In 1607 two ships under the command of Van der Hagen remained some weeks in Mauritius, and the crews feasted on an abundance of "tortoises, *dodars,* pigeons, turtles, grey parroquets, and other game." Not content with devouring numbers of these animals, it is stated that they salted quantities of tortoises and dodars for consumption during the voyage:—

"Pendant tout le temps qu'on fut là, on vêcut de tortuës, de dodarses, de pigeons, de tourterelles, de perroquets gris, et d'autre chasse, qu'on alloit prendre avec les mains dans les bois. . . . . . La chair des tortuës terrestres étoit d'un fort bon goût. On en sala, et l'on en fit fumer, dont on se trouva fort bien, demême que des dodarses qu'on sala."—Recueil des Voiages de la Compagnie des Indes Or. vol. iii. p. 195, 199. See also Prevost, Recueil des Voyages. Rouen, 1725, v. 5. p. 246.

The Dutch original is to be found in the Journal of Steven Van der Hagen in the 'Tweede deel van het begin ende voortgangh der Vereenighde Nederl. Geoctroyeerde Oostindische Compagnie,' 1646, pp. 88, 89:—

"Alle den tijt dat hier lagen, zijnde ontrent 23 dagen, aten anders niet dan Schilt-padden, Dodaersen, Duyven, &c. . . . 'T Vleesch vande Landt Schilt-padden is goet, ende smakelijck, is door eenighe van d' haeren ghesouten, ende gheroockt, dat hem wonder wel ghehouden heeft, als oock de Dodaersen, die ghesouten hebben."

7. We next come to the narrative of P. W. Verhuffen, who touched at Mauritius in 1611, and mentions Dodos under the name of *Totersten.* He describes them in nearly the same

terms as Van Neck, and adds that his sailors daily killed numbers of them for food, and that if the men were not careful the Dodos inflicted severe wounds upon their aggressors with their powerful beaks. The earliest account of this voyage is entitled Eylffter Schiffart, ander Theil, oder Kurtzer Verfolg und Continuirung der Reyse, so von den Holl-und Seeländern in die Ost Indien mit neun grossen und vier kleinen Schiffen vom 1607 biss in das 1612 Jahr, unter der Admiralschafft Peter Wilhelm Verhuffen verrichtet worden.' Published by L. Hulsius, 4to. Franckfort, 1613 :—

"Es hat auch daselbst viel Vögel als Turteltauben, grawe Papagayen, Rabos forcados, Feldhüner, Rebhüner, und andere Vögel, an der grösse den Schwanen gleich, mit grossen Köpffen, haben ein Fell, gleich einer Münchskutten über dem Kopff und keine Flügel, denn an statt derselben stehen etwan 5 oder 6 gelbe Federlein, dessgleichen haben sie auch an statt dess Schwantzes etwan 4 oder 5 uber sich gekeimte Federn stehen; von Farben seynd sie grawlecht; man nennet sie Totersten oder Walckvögel, derselben nun gibt es daselbst ein grosse menge; wie denn die Holländer täglich derselben viel gefangen und gessen haben, denn nicht allein dieselben, sondern auch ins gemein alle Vögel daselbst so zahm seyn, dass sie die Turteltauben, wie denn auch die andere wilde Tauben und Papagayen mit Stecken geschlagen, und mit den Händen gefangen haben; die Totersten oder Walckvögel haben sie mit den Händen gegriffen, musten sich aber wohl fursehen, dass sie sie nicht mit den Schnäbeln, welche sehr gross, dick und krumm seyn, etwan bey eim Arm oder Bein ergriffen, denn sie gewaltig hart zubeissen pflegen."—p. 51. See also De Bry, India Orientalis, pars ix. Supp. p. 22.

8. The figure of which the following is a fac-simile, is introduced in the Voyages of Pieter Van den Broecke, contained in the 'Begin ende voortgangh der Vereen. Nederl. Geoctr. Oost-ind. Compagnie, vol. 2, numb. xvi, p. 102.' The plate contains three figures, representing a Dodo, a single-horned Goat, and a bird not unlike the Apteryx in appearance. The goat is mentioned in the text as having been sent to the author when at Surat, as a present from the Sovereign of Agra. I can find however no notice in Van den Broecke's journal of the Dodo,

or of the other bird which he has figured, and I can therefore only conjecture that they were sketched during his visit to the Mauritius (mentioned in page 68,) which lasted from

April 19 to May 23, 1617. As the work which contains these figures is very rare, it may be well to mention that Thevenot has introduced a reversed copy of the entire plate (without stating the source) as an illustration to Bontekoe's notice of brevipennate birds in Bourbon (page 5,) to which however it can have no reference whatever.—See *Thevenot's Voyages*, vol. 1.

Though unaccompanied by any description, there can be no doubt that Van den Broecke's figure is an authentic and original representation of the Dodo, and the rudeness of the design is a proof of its genuineness. The wings are here represented as rather longer and more pointed than in the other figures.

What bird Van den Broecke's other figure may be intended to represent, or from what country it came, must be left to conjecture, and I only introduce it here from its apparently brevipennate character.

9. Sir Thomas Herbert, in 1627, visited Mauritius, and found it still uninhabited by man. In his Travels, he describes and figures the Dodo, but without adding much to our knowledge. It appears to have been the amusement of Sir T. Herbert's later days repeatedly to re-write his Travels, changing the words of each successive edition, but without much alteration in the sense. The following extracts from three editions of the work will exhibit the quaintness of the author's style, and render his observations on the Dodo more complete :—

A Relation of some yeares' Travaile, begunne Anno 1626, into Afrique and the greater Asia, especially the territories of the Persian Monarchie, and some parts of the Orientall Indies and Iles adiacent. By T. H. Esquier. Fol. London, 1634.

Some yeares Travels into divers parts of Asia and Afrique, describing especially the two famous empires the *Persian* and *Great Mogull*. Revised and enlarged by the Author. Fol. London, 1638.

"The Dodo comes first to our description: here and in *Dygarrois*,

Some Years Travels into divers parts of Africa and Asia the great. Fol. London, 1677.

"The Dodo; a bird the Dutch call Walghvogel or Dod Eersen; her body is round and fat, which occasions the slow pace, or that her corpulencie; and so great as few of them weigh

"First, here and here only and in *Dygarroys*, is generated the *Dodo*, which for shape and rarenesse may antigonize the Phœnix of *Arabia*: her body is round and fat, few weigh lesse then fifty pound, are reputed of more for wonder then food, greasie stomackes may seeke after them, but to the delicate, they are offensiue and of no nourishment.

Her visage darts forth melancholy, as sensible of Nature's injurie in framing so great a body to be guided with complementall wings, so small and impotent, that they serue only to prove her *Bird*.

The halfe of her head is naked, seeming couered with a fine vaile, her bill is crooked downwards, in midst is the thrill, from which part to the end tis of a light greene, mixt with a pale yellow tincture; her eyes are small, and like to Diamonds, round and rowling; her clothing downy feathers, her traine three small plumes, short and inproportionable, her legs suting to her body, her pounces sharpe, her appetite strong and greedy, Stones and Iron are digested, which description will better be conceiued in her representation.—P. 211.

(and no where else that ever I could see or heare of,) is generated the Dodo, (a Portuguize name it is, and has reference to her simplenes,) a Bird which for shape and rarenesse might be called a Phœnix (wer't in *Arabia*:) her body is round and extreame fat, her slow pace begets that corpulencie; few of them weigh lesse than fifty pound: better to the eye than stomack: greasie appetites may perhaps commend them, but to the indifferently curious, nourishment, but prove offensive. Let's take her picture: her visage darts forth melancholy, as sensible of Nature's injurie in framing so great and massie a body to be directed by such small and complementall wings, as are unable to hoise her from the ground, serving only to prove her a Bird; which otherwise might be doubted of: her head is variously drest, the one half hooded with downy blackish feathers; the other perfectly naked; of a whitish hue, as if a transparent Lawne had covered it: her bill is very howked, and bends downwards, the thrill or breathing place is in the midst of it; from which part to the end, the colour is a light greene mixt with a pale yellow; her eyes be round and small, and bright as Diamonds; her cloathing is of finest Downe, such as you see in Goslins: her trayne is (like a *China* beard) of three or foure short feathers; her legs thick, and black, and strong; her tallons or pounces sharp, her stomach fiery hot, so as stones and iron are easily digested in it; in that and shape, not a little resembling the *Africk* Oestriches: but so much as for their more certain difference I dare to give thee (with two others) her representation.—P. 347.

less than fifty pound: meat it is with some, but better to the eye than stomach; such as only a strong appetite can vanquish: but otherwise, through its oyliness it cannot chuse but quickly cloy and nauseate the stomach, being indeed more pleasurable to look than feed upon. It is of a melancholy visage, as sensible of Nature's injury in framing so massie a body to be directed by complemental wings, such indeed as are unable to hoise her from the ground, serving only to rank her amongst Birds: her head is variously drest, for one half is hooded with down of a dark colour; the other half naked and of a white hue, as if lawn were drawn over it; her bill hooks and bends downwards; the thrill or breathing place is in the midst; from which part to the end the colour is of a light green mixt with a pale yellow; her eyes are round and bright, and instead of feathers has a most fine down; her train (like to a *Chyna* beard) is no more than three or four short feathers: her leggs are thick and black; her tallons great; her stomach fiery, so as she can easily digest stones; in that and shape not a little resembling the Ostrich. The Dodo . . . . and one of the Hens take so well as in my table-book I could draw them."—P. 383.

Sir T. Herbert also gives a figure of what he calls "A Hen," which is very probably intended for the same bird which accompanies the Dodo in Van den Broecke's plate (*supra*, p. 19). He alludes to "Hens" among the other birds of Mauritius, but gives us no information by which

A Hen

they can now be identified. This bird is probably the same that is mentioned by Leguat, among other Mauritian birds, under the name of *Gelinottes.* The "*Velt-hoenders*" of Cornelisz (*supra*, p. 13), and the "*Feldhüner*" of Verhuffen (p. 18), may also refer to it. Compare also the words of Cauche: "Il y a en l'isle Maurice et Madagascar . . . . . . des poules rouges, *au bec de becasse;* pour les prendre il ne faut que leur presenter une pièce de drap rouge, elles suivent et se laissent prendre à la main: elles sont de la grosseur de nos poules, excellentes à manger."—Cauche, Voyage, p. 132.

10. François Cauche, in the account of his Voyage made in 1638, published in the 'Relations veritables et curieuses de l'Isle de Madagascar, Paris, 1651,' says that he saw in Mauritius birds called Oiseaux de Nazaret, larger than a swan, covered with black down, with curled feathers on the rump, and similar ones in place of wings; that the beak was large and curved, the legs scaly, the nest made of herbs heaped together, that they lay but one egg the size of a halfpenny roll, and that the young ones have a stone in the gizzard.

With a view of deducing the size of these eggs, I was contemplating an investigation of the prices of corn, the wages of labour, the honesty of bakers, and other elements, in hopes of determining the bulk of a "pain d' un sol" in 1638, but I have fortunately been spared this enquiry by another passage of Cauche, where he assigns the same dimensions to the egg of the Cape Pelican (*Pelicanus onocrotalus*), which may therefore be taken as an approximation to the size of the Dodo's egg. There can be no doubt that the bird described by Cauche was the Dodo, although his account was probably composed from memory, or confused with the descriptions then current of the Cassowary; for he tells us that it had only three toes on each foot, that the legs were of considerable length, and that the bird had no tongue, which latter character was at that time falsely attributed to the Cassowary. (See De Bry, part IV. pl. viii.) Out of this erroneous statement sprang up the "*Didus nazarenus*," a phantom-species, which has haunted our systems of ornithology from the days of Gmelin downwards. Cauche conjectures, and many authors repeat, that these birds derived their name from the island, or rather sand-bank, of Nazareth, to the north-east of Madagascar, but this idea is utterly unfounded.

Can the name *oiseau de Nazaret* have been a blunder, founded on *oiseau de nausée*, the French translation of *Walghvogel?*

We will now put Cauche himself in the witness-box :—

" J'ay veu dans l'isle Maurice des oiseaux plus gros qu'un cygne,[1] sans plumes par le corps, qui est couvert d'un duvet noir, il a le cul tout rond, le croupion orné de plumes crespues, autant en nombre que chaque oiseau a d'années, au lieu d'aisles ils ont pareilles plumes que ces dernieres, noires et recourbées, ils sont sans langues, le bec gros, se courbant un peu par dessous, hauts de jambes, qui sont escaillées, n'ayans que trois ergots à chaque pied. Il a un cry comme l'oison, il n'est du tout si savoureux à manger, que les fouches et feiques [flamingos and ducks], desquelles nous venons de parler. Ils ne font qu'un œuf, blanc, gros comme un pain d'un sol, contre lequel ils mettent une pierre blanche de la grosseur d'un œuf de poules. Ils ponnent sur de l'herbe qu'ils amassent, et font leurs nids dans les forests, si on tue le petit, on trouve une pierre grise dans son gesier, nous les appellions oiseaux de Nazaret. La graisse est excellente pour adoucir les muscles et nerfs."—Relation du Voyage de François Cauche, p. 130.[2]

11. Our next evidence is of a very important kind, as it shews that in one instance at least this extraordinary bird was brought alive to Europe, and exhibited in this country. In a MS. (Sloane MSS., 1839, 5, p. 9) in the British Museum, Sir Hamon Lestrange (the father of the more celebrated Sir Roger), in a commentary on Brown's Vulgar Errors, and *apropos* of the Ostrich, narrates as follows :—

" About 1638, as I walked London streets, I saw the picture of a strange fowle hong out upon a cloth, [hiatus in the MS.] and myselfe with one or two more then in company went in to see it. It was kept in a chamber, and was a great fowle somewhat bigger than the largest Turky Cock, and so legged and footed, but stouter and thicker and of a more erect shape, coloured before like the breast of a yong cock fesan, and on the back of dunn or deare coulour. The keeper called it a Dodo, and in the ende of a chymney in the chamber there lay a heape of large pebble stones, whereof hee gave it many in our sight, some as bigg as nutmegs, and the keeper told us shee eats them (conducing to digestion), and though I remember not how farr the keeper was questioned therein, yet I am confident that afterwards shee cast them all againe."[3]

I have endeavoured to find some confirmation from contemporary authorities of this very interesting statement, but hitherto without success. The middle of the 17th century was most prolific in pamphlets, newspapers, broadsides, "rows of dumpy quartos," and literary "rubbish-mountains," as Mr. Carlyle designates them; but the political storms of that period rendered men blind to the beauties and deaf to the harmonies of Nature, and its literature is very barren in physical research. Still there may possibly linger among our records some

[1] " La figure de cet oiseau est dans la 2 navigation des Hollandois aux Indes Orientales en la 29 diée de l'an 1598. Ils l'appellent, de nausée."

[2] " Peut-estre, que ce nom leur a esté donné, pour avoir esté trouvéz dans l'isle de Nazare, qui est plus haut que celle de Maurice, sous le 17 degré au delà l'Equateur du costé du Sud."

[3] This passage was first published in Wilkin's edition of Sir Thomas Brown's Works, 4 vols. 8vo. Lond., 1836. v. 1, p. 369; v. 2, p. 173.

black-letter hand-bill or illiterate tract, which may allude to what must have been, in that marvel-loving though unscientific age, a very attractive exhibition. To the bibliophile who shall discover such a document, I promise a splendidly-bound copy of THE DODO-BOOK. In the meanwhile we will pass on to the

12th independent notice of the Dodo, which is contained in Tradescant's Catalogue of his "Collection of Rarities preserved at South Lambeth near London," 1656. We here find one of the entries "Dodar from the island Mauritius; it is not able to flie being so big."—p. 4.

This specimen is enumerated under the head of "whole birds;" and Willughby, whose "Ornithologia" was published in 1676, speaking of the Dodo, says, "Exuvias hujusce avis vidimus in museo Tradescantiano." It is also alluded to by Llhwyd[1] in 1684, and by Hyde[2] in 1700, having meanwhile passed with the rest of Tradescant's curiosities into the Ashmolean Museum at Oxford, where the head and foot of this specimen are fortunately still extant. I shall speak further of these hereafter, and will at present only remark that this is in all probability the same individual which was exhibited in London, and which Lestrange described in 1638. Tradescant, we know, spent his life in collecting curiosities; and as there was at that time scarcely any other museum, public or private, in Great Britain to enter into competition with his, we may suppose that such a *rara avis* as this live Dodo must have been, would naturally on its decease find its way into his cabinet.[3] Another not impossible conjecture is, that this specimen was brought from Mauritius by Sir T. Herbert, who in a letter to Ashmole, quoted in Hamel's "Tradescant der Aeltere," p. 173, says, "South Lambeth, a place I well know, having been sundry times at M. Tredescon's (to whom I gave severall things I collected in my travels)." I think, however, that had the garrulous Sir Thomas actually killed, skinned, and brought home a Dodo, he would not have failed to record such an exploit in his Travels.

13. In Piso's edition of Bontius, 1658, there is a description and figure of the Dodo, though perhaps neither can be regarded as original and independent testimonies. The figure seems to be copied from one of Savery's paintings, of which I shall speak presently, and the description adds little, if anything, to the details contained in previous authors. Copies of this engraving were subsequently published in Thevenot's Voyages, vol. 1, in Willughby's Ornithology, pl. 27, and other works; but as Piso's figure is the earliest known copy from

[1] Catalogus Animalium quæ in Museo Ashmoleano conservantur; MS. No. 29.

[2] Historia Religionis veterum Persarum. 4to. Oxon. 1700, p. 312. *Apropos* of Zoroaster's mother, whose name was Dodo. He quotes Herbert's account, and adds (on what authority is unknown) that the bird laid numerous eggs, though Cauche's statement that it lays but one (confirmed by Leguat's similar assertion of the Solitaire) is more probable.

[3] Since writing the above, I see that Dr. Hamel has come to a similar conclusion.—Bull. Phys. Ac. Petersb. May 29, 1846.

Savery's designs, I have thought it worth insertion here, together with the accompanying description, which forms one of Piso's supplementary chapters to "Jacobi Bontii Historiæ naturalis et medicæ Indiæ Orientalis libri sex," contained in "Gulielmi Pisonis Medici Amstelædamensis de Indiæ utriusque re naturali et medicâ libri quatuordecim." fol. Amstelædami, 1658.

At chapter xvii., p. 70, we read:—

"De Dronte, aliis Dodaers. Inter insulas Indiæ orientalis, censetur illa quæ ab aliis *Cerne* dicitur, à nostratibus Mauritii nomen audit, ob Ebenum nigrum potissimum celebris. In hac insula frequens est miræ conformationis avis, *Dronte* dicta. Magnitudinis intra Struthionem et Gallum Indicum, à quibus ex parte figura discrepat, et ex parte cum iis convenit, imprimis cum Struthionibus Africanis, si uropygium, pennas, et plumas consideres; adeo ut Pygmæus quasi inter eos appareat, si crurum brevitatem respicias. Cæterum capite est magno, deformi, tecto quadam membrana, cucullum referente. Oculis magnis, nigris; collo curvo, prominente, pingui; rostro supra modum longo, valido, ex cœruleo albicante, exceptis extremitatibus, quarum inferior nigricat, superior flavescit, utraque acuminata, et adunca. Rictu fœdo, admodum patulo, quasi ad ingluviem nato. Corpore obeso, rotundo, quod mollibus plumis griseis, more Struthionum vestitur: ab utroque latere, loco remigum, exiguis alis plumatis, ex flavo cinereis, et pone uropygium, loco caudæ, quinis pinnulis crispis, ejusdem coloris, decoratur. Cruribus est flavescentibus crassis, sed admodum curtis, quatuor digitis pedis solidis, longis, quasi squamosis, totidem unguibus validis nigris incedit. Cæterum tardigrada est avis et stupida, quæque facile præda fit venatoribus. Caro earum, imprimis pectoris, est pinguis, vesca, adeoque multa ut tres quatuorve *Drontes* centenis sociis saturandis aliquando suffecerint. Si non probè elixentur, vel veteres sint, difficilioris sunt concoctionis, et salitæ in penu reconduntur.

Lapilli diversæ formæ et magnitudinis, cinerei coloris, in ventriculo harum avium reperiuntur, non tamen ibi nati, ut vulgus et pubes nautica arbitratur, sed in littore devorati, quasi et hoc quoque signo cum Struthionis natura aves has participare constaret, quod durissima quæque deglutiant, nec tamen digerant."

The 13th historical testimony which I have to adduce is contained in a small tract in the Ashmolean Museum (Ashm. Printed Books, No. 967). Of this there are two editions, the first without date, and entitled "A Catalogue of part of those Rarities collected in thirty years time with a great deal of Pains and Industry, by one of his Majestie's sworn Servants, R. H. alias Forges, Gentleman. They are to be seen at the place formerly called the Musique House at the West end of Pauls." Here, among other rarities, we find at p. 11, "A Dodo's Leg, it is a bird that cannot flye." The second edition is entitled, "A Catalogue of many natural rarities with great industry, cost, and thirty years travel in foraign Countries collected by Robert Hubert alias Forges, Gent. and sworn servant to his Majesty. And daily to be seen at the place formerly called the Music House near the West end of St. Paul's Church." 12mo, London, 1665. At page 11 is the following entry: "A legge of a Dodo, a great heavy bird that cannot fly; it is a Bird of the Mauricius Island." In all probability this is the same specimen that afterwards passed into the collection of the Royal Society, and is mentioned in the catalogue of their "Natural and artificial Rarities," published by Grew in 1681, who thus describes it:—

"The leg of a Dodo . . . . . . The leg here preserved is covered with a reddish yellow scale. Not much above four inches long; yet above five in thickness, or round about the joynts: wherein though it be inferior to that of an Ostrich or Cassoary, yet joyned with its shortness, may render it of almost equal strength."—p. 60.

This specimen is now preserved in the British Museum, and I shall notice it hereafter under the head of *Anatomical Evidences.*[1]

14. Olearius, in his Catalogue of the Gottorf Museum at Copenhagen, of which the first edition was published in 1666, enumerates, among other curiosities, a Dodo's head. He also gives a figure of the bird in pl. 13, f. 5, which however is merely a copy from that of Clusius (p. 12, *supra*). The following are his words:—

"Num. 5 ist ein Kopff von einem frembden Vogel welchen Clusius Gallum peregrinum, Nierenbergius Cygnum cucullatum, die Holländer aber Walghvogel, vom Eckel den sie wegen des harten Fleisches machen sollen, nennen. Die Holländer sollen zu erst solchen Vogel auff der Insel Mauritius angetroffen haben; sol auch keine Flügel, sondern an dessen Stat zwo Pinnen haben, gleich wie die Emeu und Pinguinen. Clus. exot."—Olearius, Gottorfische Kunstkammer. 4to, Schleswig. ed. of 1674.

[1] It has been supposed that this is the same leg as that described by Clusius (*supra*, p. 16), but there are certain discrepancies in the measurements which render this doubtful.

This specimen has been very recently recovered from oblivion, and is now one of the chief treasures of the Royal Museum at Copenhagen, to which I shall again refer.

15. The latest known testimony as to the existence of Dodos in Mauritius is contained in a MS. in the British Museum (Sloane MSS. 3668. Plut. cxi. F.) for a reference to which, as for many other valuable suggestions, I am indebted to J. Wolley, Esq. of Edinburgh, who has taken much interest in the history of the Dodo, and has liberally communicated the results of his researches. This document is entitled " A coppey of Mr. Benj. Harry's Journall when he was cheif mate of the Shippe Berkley Castle, Captn. Wm. Talbot then Commander, on a voyage to the Coste and Bay, 1679, which voyage they wintered at the Maurrisshes."

The Journal is little more than a ship's log, containing many rough observations, perhaps valuable, of a brilliant comet. They left Deptford 19th Nov. 1679, and on their return from India, being unable to weather the Cape of Good Hope, they determined to make for " the Marushes," the 4th June, 1681. They saw land on the 3d July, and on the 11th they began to build huts, and they had much labour in spreading their cargo out to dry :—

" After all these turmoyles, and various accidents, wee the beginning 7ber. brought all to a period: one parte of our misery wass that that time wee designed for recreation wee were forct to impt. in Labour.

" The ayre whilst wee have been here hath been very temperate neither over hott nor over cold: itt hath been showery 3 or 4 Days sucksessively, and showery in the night, sometimes a Sea Brees little wind morning and evenings.

" Now having a little respitt I will make a little description of the Island, first of its Producks then of itts parts: ffirst of winged and feathered ffowle the less passant, are *Dodos whose fflesh is very hard.* a small sort of Gees, reasonably good Teele, Curleves, Pasca fflemingos, Turtle Doves, large Batts, many small Birdes which are good.

" The Dutch pleading a propriety to the Island because of their settlement have made us pay for goates 1d per pound or ½ piece of 8 per head, the which goates are butt reasonably good, these wild, as allso the Deer which are as large as I believe any in the world, and as good fflesh in their seasons; for these 3 pie. of 8 per head, Bullocks large 6 pie. of 8 per head; [that] ys for victualling, heer are many wild hoggs and land turtle which are very good, other small creators on the Land, as Scorpions and Musketoes, these in small numbers, Ratts and ffleys a multitude, Munkeys of various sorts.

" In the woodes Eaboney, Box, Iron wood blacke and read, a false but not lasting fire, various sortes of other wood, though heavy yett good for fiering.

" In ye Sea and River, green tortoise very good, Shirkes, Doggs, Mulletts, Jackabeirs (butt nott good though some 70 lb), Breams, Pomfletts, Plaise, a ffishe like a Salmond, and heer soe called but full of small Boanes forked, severall sortes of read ffish butt nott houlsome, various sortes of small ffish for the Pann, good oysters and Crabes, Ells large and good.

" Herbage ffruite and Graine, ffrench or Cidney Beanes, Patatoes, sallating; Pumplemuses, oranges, Jumboes, Watter and musk Melones, Sugar Cannes, Pumkines, Tobacco that Hellish weed, and many other things forgotten."

Such then is the sum of the *Historical Evidence* which we possess for the existence of this singular creature. In 1644 the Dutch first colonized the island of Mauritius, and it is probable that these gigantic fowls, deprived of flight, slow of foot, and useful for food, were speedily diminished in number, and finally exterminated by the thoughtless rapacity of the early colonists. Their destruction would be further hastened, or might be mainly caused, by the Dogs, Cats, and Swine which accompany Man in his migrations, and are speedily naturalized in the forests. To such animals the eggs and young of the Dodo and other birds would be a dainty treat; and that this is no mere conjecture is proved by Leguat, who tells us, "Here, (in Mauritius,) are Hogs of the China kind. . . . . These beasts do a great deal of damage to the inhabitants, by devouring all the young animals they can catch."—p. 170, Eng. ed.

That the destruction of the Dodos was completed by 1693, may be inferred from the narrative of Leguat, who in that year remained several months in Mauritius, and enumerates its animal productions at some length, but makes no mention whatever of Dodos. He further says, "L'isle etait autrefois toute remplie d'Oyes et de Canards sauvages; de Poules d'eau, de Gelinottes, de Tortues de mer et de terre; *mais tout cela est devenu fort rare.*" This passage proves, that even in 1693, civilization had made great inroads on the fauna of Mauritius.

In 1712 the Dutch evacuated Mauritius, and the French colonized the island under the new name of Isle de France. This change in the population will account for the absence of any traditionary knowledge of so remarkable a bird among the later inhabitants. All subsequent evidence is equally negative. Baron Grant resided in Mauritius from 1740 to 1760; and his son, who compiled the "History of Mauritius" from his papers, states (p. 145*) that no trace of such a bird was to be found at that time. M. Morel, a French official who resided there previously to 1778, and whose attention seems to have been drawn to the subject by the judicious criticisms of Buffon (Hist. Ois. vol. ii. p. 73), tells us that the oldest inhabitants had no recollection of these creatures (Observations sur la Physique, 1778, vol. xii. p. 154). The late M. Bory de St. Vincent remained for some time in Mauritius and Bourbon in 1801, and has left an excellent work on the physical features of those islands (Voyage dans les quatre principales iles des Mers d'Afrique). He assures us (vol. ii. p. 306) that he made every possible enquiry respecting the Dodo and its allies, without gaining the slightest information from the inhabitants on the subject. At a public dinner at the Mauritius in 1816, several persons from 70 to 90 years of age were present, who had no knowledge of such a bird from recollection or tradition (De Blainville in Nouv. Ann. Mus. vol. iv. p. 31). Mr. J. V. Thompson also resided for some years in Mauritius and Madagascar, previously to 1816, and he states that no more traces of the existence of the Dodo could then be found, than of the truth of the tale of Paul and Virginia, although a very general idea prevailed as to the reality of both (Mag. Nat. Hist. ser. 1, vol. ii. p. 443). This list of negative witnesses may be closed with the late Mr. Telfair, a very active naturalist, whose researches were equally conclusive as to the non-existence of Dodos in Mauritius in modern times (Zool. Journ. vol. iii. p. 566).

SECTION II.—*Pictorial Evidences—Picture in the British Museum—Roland Savery's picture at the Hague; his picture at Berlin; his picture at Vienna—John Savery's picture at Oxford.*

THE next series of evidences to be adduced are those derived from contemporary paintings. We have seen that the narratives of the early voyagers are in several instances accompanied by rude delineations of Dodos, but besides these we possess certain oil paintings of this bird by artists of great merit, who apparently aimed only at correctly representing the object before them. All these pictures, except one, closely resemble each other, and though exhibiting slight variations, they seem to have been taken from one original design. They moreover agree sufficiently well with the engravings in the early voyages, to leave no doubt of their being intended for the same species of bird. Five of these paintings are now known to exist; one of these is anonymous, three bear the name of Roland Savery, an eminent Dutch animal painter in the beginning of the 17th century, and one is by John Savery, the nephew of Roland.

1. The first of these paintings, and the best known, is that from which the figure of the Dodo in all modern books of natural history has been copied. This picture was once the

property of the artist, George Edwards, who in his work on Birds, vol. vi, pl. 294, tells us, "The original picture was drawn in Holland from the living bird, brought from St. Maurice's

Island in the East Indies, in the early times of the discovery of the Indies by the way of the Cape of Good Hope. It was the property of the late Sir H. Sloane to the time of his death, and afterwards becoming my property I deposited it in the British Museum as a great curiosity. The above history of the picture I had from Sir H. Sloane and the late Dr. Mortimer, secretary to the Royal Society." This picture is still preserved in the British Museum, and may be seen in the Bird Gallery along with the Dodo's foot, to be hereafter described. It represents the Dodo surrounded by American Maccaws, Ducks, and other birds, depicted with great exactness and attention to details. Judging from the animated and natural expression which the artist has introduced, I am quite disposed to believe the assertion of Edwards, that it was painted from life. Unfortunately there is neither name nor date upon the picture; but from the style of execution, and the identity of the design with the pictures next to be noticed, it may be attributed to one of the two Saverys. As the other birds in this picture are the size of life, the Dodo is probably represented of its true magnitude, although it must have been a rather larger specimen than either of those whose skulls are now extant.

The engraving on the opposite page was made under Mr. Broderip's superintendance, to illustrate his treatise in the Penny Cyclopædia, and as it is an accurately reduced copy of the painting in question, I have obtained the permission of Messrs. Clowes to introduce it here.

2. In the Royal Collection at the Hague is a painting by Roland Savery, which is pronounced by Houbracken (Groote Schouburgh der Nederlantsche Konstschilders en Schilderessen, Hague, 1753, vol. i. p. 58,) to be one of that master's *chef d'œuvres*.[1] It represents Orpheus charming the animal creation with his music, and among innumerable birds and beasts, which are depicted with the utmost accuracy, we see the clumsy Dodo spell-bound by the strains of the Lyric Bard. All the other animals in this composition are exact and almost mechanical copies of nature, without the smallest indication of pictorial licence; we cannot therefore suppose that the artist would have marred the consistency of his design by introducing a fabulous or even an exaggerated representation. The Dodo, like all the other figures, must have been copied from careful sketches made either by the artist himself or by persons in whom he could confide. Such were my own impressions on examining this painting in 1845, and Professor Owen, who was the first to call the attention of Naturalists to it, expresses a similar opinion.

"Whilst at the Hague," he says, "in the summer of 1838, I was much struck with the minuteness and accuracy with which the exotic species of animals had been painted by Savery and Breughel, in such subjects as Orpheus charming the beasts, &c., in which scope was allowed for grouping together a great variety of animals. Understanding that the celebrated menagerie of Prince Maurice had afforded the living models to these artists, I sat down one day before Savery's Orpheus and the Beasts, to make a

[1] Dr. Hamel, in his recently published work entitled 'Tradescant der Aeltere,' p. 170, states that this picture was painted in 1638, but he has probably no other authority than the conjecture that the bird shewn that year in London served as Savery's model.

list of the species which the picture sufficiently evinced that the artist had had the opportunity to study alive. Judge of my surprise and pleasure in detecting in a dark corner of the picture (which is badly hung between two windows) the *Dodo*, beautifully finished, showing for example, though but three inches long, the auricular circle of feathers, the scutation of the tarsi, and the loose structure of the caudal plumes. In the number and proportions of the toes, and in general form, it accords with Edwards's oil painting in the British Museum; and I conclude that the miniature must have been copied from the study of a living bird, which, it is most probable, formed part of the Mauritian menagerie. The bird is standing in profile with a lizard at its feet. Not any of the Dutch naturalists to whom I applied for information respecting the picture, the artist, and his subject, seemed to be aware of the existence of this evidence of the Dodo in the Hague collection."—Penny Cyclopædia, vol. xxiii. p. 143.

3. Shortly after visiting the Hague in 1845, I made a search in the Royal Gallery at Berlin, which contains several of Roland Savery's highly finished paintings. Among them I found one which represents numerous animals in Paradise, one of which is a Dodo, of about the same size and in nearly the same attitude as the figure last mentioned. But what renders this picture peculiarly interesting is, that it affords us a date, the words "Roelandt Savery fe. 1626," being painted in one corner. (See Frontispiece.) As Roland Savery was born in 1576, he was 23 years old when Van Neck's expedition returned to Holland; and as we are told by De Bry that the Dutch brought home a Dodo on that occasion, it is possible enough that Savery may have taken the portrait of this individual, and that the design thus made may have been copied by himself and by his nephew John in their later pictures. Or if we feel disposed (for the reasons given at p. 11, *supra*) to doubt the correctness of De Bry's statement, we may yet suppose, with Professor Owen, that the menagerie of Prince Maurice supplied the living prototype for Savery's pencil. This opinion is corroborated by the tradition recorded by Edwards, that the picture in the British Museum was drawn in Holland from the living bird. It is far more probable than the conjecture of Dr. Hamel, (Bull. Ac. Petersb. vol. v. p. 317) that Savery's pictures were copied from the Dodo exhibited in London, as this individual must in that case have lived in captivity at least 12 years, from 1626 to 1638.

4. The present sheet was just rescued from the printer in time to announce an important addition to our Pictorial Evidence. Dr. J. J. de Tschudi, the eminent Peruvian traveller, hearing that this work was in preparation, has had the kindness to transmit to me an exact copy of a figure of the Dodo by Roland Savery, which forms part of a picture in the imperial collection of the Bellvedere at Vienna, and which is here introduced. (Plate III.) Dr. Tschudi states that this picture is dated 1628; two years later than the Berlin one. There are two circumstances which give an especial interest to this painting. First, the novelty of attitude in the Dodo, exhibiting an activity of character which corroborates the supposition that the artist had a living model before him, and contrasting strongly with the aspect of passive stolidity in the other pictures. And, secondly, the Dodo is represented as watching, apparently with hungry looks, the merry wrigglings of an eel in the water! Are we hence to infer that the Dodo fed upon eels? The advocates of the Raptorial affinities of the Dodo,

Fac-simile of Savery's picture of the DODO, in the Bellvedere at Vienna.

of whom we shall soon speak, will doubtless reply in the affirmative, but as I hope shortly to demonstrate that it belongs to a family of birds, all the other members of which are frugivorous, I can only regard the introduction of the eel as a pictorial license. In this, as in all his other paintings, Savery brought into juxta-position animals from all countries, without regarding geographical distribution. His delineations of birds and beasts were wonderfully exact, but his knowledge of natural history probably went no further, and although the Dodo is certainly *looking at* the eel, yet we have no proof that he is going to *eat* it. The mere collocation of animals in an artistic composition, cannot be accepted as evidence against the positive truths revealed by Comparative Anatomy.

5. The last painting which I have to mention is the one presented to the Ashmolean Museum at Oxford in 1813, by W. H. Darby, Esq., but of whose previous history nothing is known. It is painted by John Savery, the nephew of Roland, and bears the date of 1651. It appears to be copied from the same original design as the three first pictures above referred to, but a remarkable feature in it is its colossal scale, the Dodo standing about 3 feet 6 inches high, and being nearly double the size which the picture in the British Museum, the description of eye-witnesses, and the existing remains warrant us in attributing to the bird. It is difficult to assign a motive to the artist for thus magnifying an object already sufficiently uncouth in appearance. Were it not for the discrepancy of dates, I should have conjectured that this was the identical "picture of a strange fowle hong out upon a cloth," which attracted the notice of Sir Hamon Lestrange and his friends as they "walked London streets" in 1638; the delineations used by showmen being in general more remarkable for attractiveness than veracity.

SECTION III.—*Real or Anatomical evidences—Dodo's foot in British Museum—Head and foot at Oxford—Head at Copenhagen—Probability of finding further remains in Mauritius—Figure of Dodo as deduced from evidence—Non-development of certain organs no proof of imperfection.*

I COME lastly to speak of the evidence afforded us by the few imperfect remains of this extraordinary bird which have come down to us. Portions of probably three distinct individuals of the Dodo are now extant in as many public museums. It is remarkable, as proving the interest which the discovery of the Dodo excited in Europe, that each of these three specimens is specially referred to in museum catalogues printed in the 17th century.

1. The first of these is the Dodo's leg, or rather foot, mentioned, as before stated, by Hubert in 1665, and by Grew in 1681. From the cabinet of the Royal Society it was transferred early in the last century to the British Museum, where it is now preserved.[1] It appears

[1] M. de Blainville inadvertently states, that "this leg passed into the British Museum at the end of the last or beginning of the present century, when the Museum was established through the influence of Sir J. Banks." (Nouv. Ann. Mus. H. N. vol. iv. p. 15). A little more attention to names, dates, and such minutiæ, would have added to the value of this memoir.

to have attracted no further notice till 1793, when Dr. Shaw gave a figure of it in his Naturalists' Miscellany, pl. 143. This foot seems to have belonged to a somewhat larger individual than the Ashmolean specimen, and from its excellent preservation exhibits the external characters of the tarsus and toes in a very interesting manner. (See Plate VI.)

2. The stuffed specimen of the Dodo mentioned in the catalogue of Tradescant's Museum, 1656, was bequeathed with the rest of his curiosities to Elias Ashmole, the munificent founder of the Ashmolean Museum at Oxford. Here it remained in an entire, if not a very perfect state, till 1755, when the Vice-Chancellor and the other Trustees, to whose guardianship the worthy Ashmole had confided his treasures, came in an unlucky hour to make their annual visitation of the Museum. In those days Oxford presented the still existing anomaly of a University, in which Zoology was not publicly taught as a science; the Royal Society had long removed to the metropolis, the Ashmolean Society was as yet unborn, and the Taylor Institution had not opened a door to continental literature. The literary and scientific ardour which Lister, Plott, Aubrey, Ashmole, Wood, Llhwyd, and others had awakened in the 17th century had now subsided, and the University seems to have relapsed into the scholastic torpor of the middle ages. We need not therefore wonder at the fate which befel the LAST OF THE DODOS. This unhappy specimen, then at least a century old, had, it appears, become decayed by time and neglect; and according to a record now extant, was, with many others, "ordered to be removed at a meeting of a majority of the visitors."[2] On this fatal decree, Mr. Lyell appropriately remarks (and Mr. Broderip will forgive my re-quoting the passage):—

> "Some have complained that inscriptions on tombstones convey no general information except that individuals were born and died—accidents which happen alike to all men. But the death of a *species* is so remarkable an event in natural history, that it deserves commemoration; and it is with no small interest that we learn from the archives of the University of Oxford the exact day and year when the remains of the last specimen of the Dodo, which had been permitted to rot in the Ashmolean Museum, were cast away. The relics we are told were "a Museo subducta, annuentibus Vice-cancellario aliisque Curatoribus, ad ea lustranda convocatis, die Januarii 8vo, A.D. 1755."

By a lucky accident, however, a small portion of this last descendant of an ancient race escaped the clutches of the destroyer. The head and one of the feet were saved from the flames, and are still preserved in the Ashmolean Museum. The head is figured in Plate V., and is in tolerable preservation, exhibiting the remarkable form of the beak and nostrils, the bare skin of the face, and the partially feathered occiput which the old authors compared to a hood. The eyes still remain dried within the sockets, but the corneous extremity of the beak has perished, so that it scarcely exhibits that strongly hooked termination so conspicuous in all the original portraits. The deep transverse grooves are also visible, though less developed than in the paintings, from which, and from its inferior size we may infer it to have been a female specimen. The scientific value of this specimen has lately been

[2] For particulars of this act of well meant, but too sweeping, reform, see Mr. J. Duncan's paper in the Zoological Journal, vol. iii. p. 559.

very greatly increased by the careful dissection which Dr. Acland, the Lecturer in Anatomy, has made of one side of the cranium.[1] By dividing the skin down the mesial line, and removing it from the left side, the entire osteological structure of this extraordinary skull is exposed to view, while on the other side of the head the external covering remains undisturbed. See Plates VIII. and IX.

The foot, which accompanies this interesting cranium, was formerly covered with decomposed integuments, which presented few external characters. These have recently been removed by Dr. Kidd, the Professor of Medicine, who has made an interesting preparation of the osseous and tendinous structures, and exhibited some remarkable characters to which I shall presently advert.

3. I have now to speak of the cranium, mentioned by Olearius as being, in 1666, in the Gottorf Museum at Copenhagen. This specimen, after being forgotten for nearly two centuries, was very lately discovered by Professor C. Reinhardt (see Kröyer's Tidskrift, vol. iv. p. 71, and Lehmann in Nov. Act. Ac. Leop. Car. vol. xxi. p. 491), amongst a heap of venerable rubbish, and is now in the public museum at Copenhagen, where, two years ago, I had an opportunity of examining it. All the soft parts are removed, and it exhibits the same important osteological characters which have been recently brought to light in the Oxford head. It is, however, less perfect, the base of the occiput being removed. It is about half an inch shorter than the Ashmolean specimen, and proportionably smaller.

These are the only known fragments which are ascertained to be genuine relics of the Dodo. Yet it cannot be doubted that if a judicious series of researches were made in the caves and superficial deposits of the island of Mauritius, many more osseous remains might be disinterred, and possibly the entire skeleton might be reconstructed. I rejoice to find, by a recent letter from G. C. Cuninghame, Esq. to Sir W. C. Trevelyan, that this problem has attracted the attention of the Natural History Society of Mauritius, who propose making excavations for this especial object.

Let us now endeavour to combine into one view the results of the historical, pictorial, and anatomical data which we possess respecting the Dodo. We must figure it to ourselves as a massive clumsy bird, ungraceful in its form, and with a slow waddling motion. We cannot form a better idea of it than by imagining a young Duck or Gosling enlarged to the dimensions of a Swan. It affords one of those cases, of which we have many examples in Zoology, where a species, or a part of the organs in a species, remains permanently in an undeveloped or infantine state. Such a condition has reference to peculiarities in the mode of life of the animal, which render certain organs unnecessary, and they therefore are retained through life

[1] Zoologists are indebted to P. B. Duncan, Esq., Keeper of the Ashmolean Museum, who liberally permitted this important dissection of a unique specimen to take place, and I have great pleasure also in recording that it was performed " annuentibus Vice-cancellario aliisque Curatoribus," A.D. 1847.

in an imperfect state, instead of attaining that fully developed condition which marks the mature age of the generality of animals. The Greenland Whale, for instance, may be called a *permanent suckling;* having no occasion for teeth, the teeth never penetrate the gums, though in youth they are distinctly traceable in the dental groove of the jaws. The Proteus, again, is a *permanent tadpole;* destined to inhabit the waters which fill subterranean caverns, the gills which in other Batrachian Reptiles are cast off as the animal approaches maturity, are here retained through life, while the eyes are mere subcutaneous specks, incapable of contributing to the sense of vision. And lastly (not to multiply examples), the Dodo is (or rather was) a *permanent nestling,* clothed with down instead of feathers, and with the wings and tail so short and feeble, as to be utterly unsubservient to flight.[1]

It may appear at first sight difficult to account for the presence of organs which are practically useless. Why, it may be asked, does the Whale possess the germs of teeth which are never used for mastication? Why has the Proteus eyes when he is especially created to dwell in darkness? and why was the Dodo endowed with wings at all, when those wings were useless for locomotion.? This question is too wide and too deep to plunge into at present; I will merely observe, that these apparently anomalous facts are really the indications of laws which the Creator has been pleased to follow in the construction of organized beings; they are inscriptions in an unknown hieroglyphic, which we are quite sure mean *something,* but of which we have scarcely begun to master the alphabet. There appear, however, reasonable grounds for believing that the Creator has assigned to each class of animals a definite type or structure from which He has never departed, even in the most exceptional or eccentric modifications of form. Thus, if we suppose, for instance, that the abstract idea of a Mammal implied the presence of teeth, the idea of a Vertebrate the presence of eyes, and the idea of a Bird the presence of wings, we may then comprehend why in the Whale, the Proteus, and the Dodo, these organs are merely *suppressed,* and not wholly *annihilated.*

And let us beware of attributing anything like *imperfection* to these anomalous organisms, however deficient they may be in those complicated structures which we so much admire in other creatures. Each animal and plant has received its peculiar organization for the purpose, not of exciting the admiration of other beings, but of sustaining its own existence. Its perfection, therefore, consists, not in the number or complication of its organs, but in the adaptation of its whole structure to the external circumstances in which it is destined to live. And in this point of view we shall find that every department of the organic creation is equally perfect; the humblest animalcule, or the simplest conferva, being as completely organized with reference to its appropriate habitat, and its destined functions, as Man himself, who claims to be lord of all. Such a view of the creation is surely more philosophical than

[1] Our efforts to *realize* this extinct creature will be assisted by the skill of Mr. Jenssen, a sculptor at Copenhagen, who has made, or is making, plaster casts of the Dodo, the size of life, and coloured from Savery's pictures (Bull. Phys. Ac. Petersb. vol. v. p 318). We may hope that some examples of this work of art will soon reach Britain.

the crude and profane ideas entertained by Buffon and his disciples, one of whom calls the Dodo "un oiseau bizarre, dont toutes les parties portaient le caractère d'une conception manquée." He fancies that this imperfection was the result of the youthful impatience of the newly-formed volcanic islands which gave birth to the Dodo, and implies that a steady old continent would have produced a much better article (Bory St. Vincent, Voy. aux Isles des Mers d'Afrique. vol. ii. p. 305. vol. iii. p. 169).

Section IV.—*Affinities of the Dodo—Not allied to the Grallatorial or Natatorial orders; nor to the Rasores; nor to the Raptores—Opinions of Vigors; of De Blainville; of La Fresnaye; of Gould; of Gray; of Broderip; of Owen—Affinity of the Dodo to the Pigeons, proved by numerous agreements of structure.*

We now approach the most difficult part of our subject, viz., to determine, from such imperfect data as history and anatomy present, the affinities of the Dodo to other generic forms in the class of Birds. Now it is evident, at first sight, that the Dodo is a very anomalous and exceptional animal; in the language of systematists, it forms a very isolated genus, far removed from the large groups in which the more prevalent arrangements of ornithic structure are displayed; just as its native island is intermediate between Asia and Africa, and can hardly be referred to either continent. We must not, therefore, expect to discover any very close or satisfactory affinities between the Dodo and other birds. All that we can do is to seek for those other generic forms to which, in the majority, or rather the preponderance of its characters, it makes the nearest approach.

The most prominent characteristic of the Dodo is manifestly its inability to fly, in consequence of the shortness of its wings. This is an exceptional peculiarity which occurs in only three families of existing birds,—the Penguins, the Auks, and the Ostriches. It is, therefore, natural to inquire whether the imperfectly developed wings of the Dodo indicate an affinity to any of these families. Now the Penguins are the most completely aquatic of all birds, their feathers are almost reduced into the condition of scales, and their wings are practically converted into fins, while the palmated and plantigrade feet at once prove their entire disconnection from the type of the Dodo. The Auks, of which a single species, *Alca impennis*, has the wings too short for flight, while the other species of the group are volatile, represent geographically in the northern hemisphere, the Penguins of the southern, and are equally remote from the bird before us. The Struthious birds make a somewhat nearer approach to the Dodo, in the rudimental nature of their plumage, but their long legs and neck, the comparatively feeble beak, the absence, or very slight development, of the hallux, and numerous other peculiarities, prove them to be modifications of the Grallatorial order, and by no means nearly allied to the Dodo. The apparently similar texture of plumage in the Ostriches and the Dodo (so far as we are acquainted with the latter), does not necessarily indicate any affinity; for a terrestrial bird of whatever order, if deprived of the means of flight, would, of course, have its feathers so modified as to serve only as a clothing to the skin, and they would no longer exhibit that peculiar compactness, and those beautiful mechanical arrangements which are seen in the feathers of volatile birds.

If the Dodo, then, be neither a Penguin, an Auk, nor an Ostrich, it must evidently be either an entirely unique and independent organization, representing in its own person a whole order of birds, or (which is far more probable) it must be an exceptional form of some other group, to which it stands in the same relation as the Ostriches to the Bustards, the Penguins to the Divers, or the *Alca impennis* to the other genera of *Alcidæ*.

We have seen that the Dodo can be referred neither to the Grallatorial nor Natatorial orders. Its great bulk, and the vast strength and curvature of the beak, seem equally to remove it from the Insessores, properly so called. There apparently remain, therefore, only the Gallinaceous and Raptorial orders with which we can compare it.

Before stating my own views of this question, I will give a brief notice of the opinions of some recent naturalists, whose criticisms are philosophical in spirit, if not correct in result. The arrangements of earlier systematists may be omitted, as being too crude and vague to be worth recording.

Mr. Vigors, in his elaborate paper on the "Affinities of Birds," in the Linnæan Transactions,[1] vol. xiv. p. 484, referred the Dodo to the Gallinaceous order, and considered it to be intermediate between the *Struthionidæ* and the genus *Crax*. His words are as follows:—

"The bird in question, from every account which we have of its economy, and from the appearance of its head and foot, is decidedly gallinaceous; and from the insufficiency of its wings for the purposes of flight, it may with equal certainty be pronounced to be of the *Struthious* structure, and referable to the present family (*Struthionidæ*). But the foot has a strong hind toe, and, with the exception of its being more robust,—in which character it still adheres to the *Struthionidæ*,—it corresponds exactly with the foot of the Linnæan genus *Crax*, that commences the succeeding family. The bird thus becomes osculant, and forms a strong point of junction between these two conterminous groups, which though evidently approaching each other in general points of similitude, would not exhibit that intimate bond of connection which we have seen to prevail almost uniformly throughout the neighbouring subdivisions of nature, were it not for the intervention of this important genus."

M. De Blainville, in the Nouvelles Annales du Muséum d'Histoire Naturelle, vol. iv. p. 24, objects to this arrangement on the following grounds: 1st, the form of the beak, in which the strength, the terminal hook, the nudity of the base, the width of the gape, remind us (as he says) of a rapacious rather than of a granivorous bird; 2ndly, the position of the nostrils, which are not provided with an incumbent scale; 3rdly, the strength and curvature of the claws; 4thly, the strength and shortness of the legs; 5thly, the squamous covering of the tarsi; 6thly, the short and woolly plumage of the head and neck; 7thly, the alleged toughness and bad taste of the flesh; and 8thly, the absence of metatarsal spines. He consequently concludes

[1] M. De Blainville, who seems to be acquainted with this valuable paper by Mr. Vigors, only from a brief notice of it in Mr. Duncan's "Memoir on the Dodo," in the Zoological Journal, vol. iii. p. 558, tells us that it is written by "un auteur anonyme, mais que je crois être M. Macleay."

that the Dodo is a Raptorial bird, allied to the Vultures, in proof of which he adduces: 1, the eyes placed in the smooth area of the beak, as in *Cathartes*; 2, the oval nostrils placed very forward on the beak, and without incumbent scale; 3, the form, size, and colour of the beak, resembling those of *Sarcorhamphus*; 4, the form of the cranium, its width between the orbits, its flattening on the sinciput, as in the last-named Vulture; 5, the two caruncular folds at the base of the curved portion of the beak, somewhat as in *Sarcorhamphus*; 6, the hood of skin like that of *Cathartes*; 7, the almost naked neck, of a greenish colour; 8, the form, number, and arrangement of the toes, and the strength and curvature of the claws; 9, the squamose system of the tarsi and toes; 10, the crop at the base of the neck and the muscular stomach, which are common, as he says, to the two orders; and 11, the absence of the metatarsal spine.

Notwithstanding these apparent agreements with the Rapacious order, M. de Blainville admits that the legs of the Dodo are much shorter and stronger than in any known Vulture; that the toes are not connected, as in the Vultures, by a membrane; and that the inability to fly appears even a greater anomaly in a rapacious, than in a gallinaceous bird. These difficulties, however, do not prevent him from giving his vote in favour of the Raptorial affinities of the Dodo.

The Baron de la Fresnaye, in an outline of his classification of the Birds of Prey, adopts M. de Blainville's views, and makes the *Didinæ* the first, or lowest, sub-family of the *Vulturidæ* (Revue Zoologique, 1839, p. 193). In accordance with this idea, he conjectures that the Dodo inhabited the sea-coasts, and fed upon the remains of Crustacea, Mollusca, and other offal cast up by the waves.

Mr. Gould, from a consideration of the several characters above enumerated, and especially the compression of the beak and nudity of the face, arrived at the same conclusion as M. de Blainville (Nouv. Ann. Mus. Hist. Nat. vol. iv. p. 34).

Mr. J. E. Gray has expressed the opinion that the bird represented in the pictures of the Dodo was made up artificially by joining the head of a bird of prey approaching the Vultures, if not belonging to that family, to the legs of a Gallinaceous bird. But, as Mr. Broderip well remarks, "if this be granted, see what we have to deal with. We have then two species, which are either extinct, or have escaped the researches of all zoologists, to account for; one, a bird of prey, to judge from its bill, larger than the Condor; the other, a Gallinaceous bird, whose pillar-like legs must have supported an enormous body." Mr. Gray's opinion is based on the following grounds:—

"1. The base of the bill is enveloped in a *cere*, as may be seen in the cast, where the folds of the *cere* are distinctly exhibited, especially over the back of the nostrils. The *cere* is only found in the Raptorial birds.

"2. The nostrils are placed exactly in front of the *cere*, as they are in the other *Raptores*; they are oval and nearly erect, as they are in the *true Vultures*, and in that genus alone, and not longitudinal as they are in the *Cathartes*, all the *Gallinaceous birds*, *Grallatores*, and *Natatores*; and they are naked, and covered with an arched scale, as is the case in all the *Gallinaceæ*.

"3. In Edwards's picture the bill is represented as much hooked (like the *Raptores*) at the tip; a character which unfortunately cannot be verified on the Oxford head, as that specimen is destitute of the horny sheath of the bill, and only shows the form of the bony core.

"With regard to the size of the bill, it is to be observed that this part varies greatly in the different species of Vultures, indeed so much so that there is no reason to believe that the bird of the Oxford head was much larger than some of the known Vultures.

"With regard to the foot, it has all the character of that of the *Gallinaceous birds*, and differs from all the Vultures in the shortness of the middle toe, the form of the leg, and the bluntness of the claws." (Penny Cyclopædia, vol. ix. p. 55.)

Mr. Broderip, on the other hand, after a full discussion of the question, sums it up as follows:—

"If the picture in the British Museum and the cut in Bontius be faithful representations of a creature then living, to make such a bird a bird of prey—a Vulture, in the ordinary acceptation of the term—would be to set all the usual laws of adaptation at defiance. A Vulture without wings! How was it to be fed? And not only without wings, but necessarily slow and heavy in progression on its clumsy feet. The *Vulturidæ* are, as we know, among the most active agents for removing the decomposing animal remains in tropical and intertropical climates, and they are provided with a prodigal development of wing to waft them speedily to the spot tainted by the corrupt incumbrance. But no such powers of wing would be required by a bird appointed to clear away the decaying and decomposing masses of a luxuriant tropical vegetation—a kind of Vulture for vegetable impurities, so to speak,—and such an office would not be by any means inconsistent with comparative slowness of pedestrian motion."

Professor Owen has lately made a more minute examination of the remains preserved at Oxford than was in the power of M. de Blainville, who was only acquainted with these relics through the medium of drawings and casts. The former was further aided by the recent dissection of the foot, made by Dr. Kidd, and has given us the result of his observations in a memoir published in 1845, in the Transactions of the Zoological Society, vol. iii. p. 331. Mr. Owen remarks, that the Dodo differs from all Raptorial birds "in the greater elevation of the frontal bones above the cerebral hemispheres, in the sudden sinking of the interorbital and nasal region of the forehead, in the rapid compression of the beak anterior to the orbits, in the elongation of the compressed mandibles, and in the depth and direction of the sloping symphysis of the lower jaw." He further adds that the eyes are smaller in proportion, and the nostrils more in advance and lower down than in the *Vulturidæ*.

The arguments adduced by Professor Owen in favour of its affinity to the Vultures, from a comparison of the bones of the foot with those of the common Cock, *Crax*, and other *Gallinæ*, on the one hand, and of the Vulture and Eagle on the other, will be stated at length in Part II. of this work. He concludes as follows:—

"Upon the whole, then, the Raptorial character prevails most in the structure of the foot, as in the general form of the beak of the Dodo, and the present limited amount of our anatomical knowledge of the extinct terrestial bird of the Mauritius supports the conclusion that it is an extremely modified form of the Raptorial order. Devoid of the power of flight, it could have had small chance of obtaining food by preying upon the members of its own class; and if it did not exclusively subsist on dead and decaying organized matter, it most probably restricted its attacks to the class of Reptiles and to the littoral Fishes, Crustacea, &c., which its well-developed back-toe and claw would enable it to seize and hold with a firm gripe."

It is however evident from the many counter-arguments which both De Blainville and Owen have with great impartiality adduced, that their conclusions as to the Raptorial affinities of the Dodo are far from being absolutely demonstrated. If there are objections to the Gallinaceous hypothesis, there are at least as many to the Raptorial one, and the systematic zoologist finds no more satisfaction in the one conclusion than in the other. If however we look a little further into the field of ornithic creation, we shall find a family of birds ready to claim relationship with this pedestrian outcast, and to admit him among their kindred.

The various zoologists who have hitherto attempted the classification of the Dodo, appear to have been unconsciously influenced by its colossal stature, and they consequently compared it only with birds of large size, like the Ostrich, the Vulture, or the Albatross. But although each zoological group is characterized by certain limits of magnitude, yet the range between those limits is often very great, and where the characters of structure in two organisms essentially correspond, no amount of diversity in mere size ought to justify their separation. It is by overcoming this prejudice, as to the importance of size in classification, that the *Menura*, e. g., has been recently removed from the *Rasores* to its true place among the *Insessores*, and I must now call upon zoologists to make a similar concession in regard to the Dodo.

The extensive group of *Columbidæ*, or Pigeons, is very isolated in character, and though probably intermediate between the Insessorial and Gallinaceous orders, can with difficulty be referred to either. In this group we find some genera that live wholly in trees, others which are entirely terrestrial, while the majority, of which the common Wood-Pigeon is an instance, combine both these modes of life. But the main characteristic of all is their diet, composed almost exclusively of the seeds of various plants and trees. We accordingly find much diversity in the forms of their beaks, according to the size and mechanical structure of the seeds on which each genus is destined to live. Those which feed on cereal grains and the seeds of small grasses and other plants, like the Common Pigeon and Turtle-dove, have the beak considerably elongated, feeble, and slender. But in tropical countries there are several groups of Pigeons called Nutmeg-eaters and Trerons, which feed on the large fruits and berries of various kinds of palms, fig, nutmeg, and other trees. These birds, and especially those of the genus *Treron* (*Vinago* of Cuvier), have the beak much stouter than other Pigeons, the corneous portion being strongly arched and compressed, so as greatly to resemble the structure of certain Rapacious birds, especially of the Vulturine family.

This Raptorial form of beak is carried to the greatest extent in the genus *Didunculus*, a very singular bird of the Samoan Islands in the Pacific Ocean (see plate VII. f. 1). Very little is yet known of its habits, but Mr. Stair, a missionary recently returned from those islands, has reported that the bird feeds upon bulbous roots. Its first discoverer, Mr. Titian Peale, an American naturalist (whose account is, I believe, still unpublished), saw something in its form or habits that reminded him of the Dodo, and hence its generic name. Sir W. Jardine, who first described the bird, under the name of *Gnathodon strigirostris*, in the Annals of Natural History, vol. xvi. p. 175, referred it conjecturally to the *Megapodidæ*, though he recognised in it several dove-like characters. And Mr. Gould, who has given two figures of it in his Birds of Australia, Part 22, pronounces that the bird approaches nearest to the Pigeons. We shall soon see that the *Didine* and *Columbine* hypotheses, though apparently incongruous, resolve themselves (as often happens) into one Truth.

Although certain genera of *Columbidæ* are thus seen to assume a form of beak resembling that of the *Raptores*, yet no two groups in the same class can be more opposed in habits and affinities than the "feroces Aquilæ" and "imbelles Columbæ." It is interesting, however, to observe that mechanical strength, whether for the devouring of animal or vegetable substances, is obtained in both cases by a similarity of structure.

If now we regard the Dodo as an extreme modification, not of the Vultures, but of these Vulture-like frugivorous pigeons, we shall, I think, class it in a group whose characters are far more consistent with what we know of its structure and habits. There is no *a priori* reason why a Pigeon should not be so modified, in conformity with external circumstances, as to be incapable of flight, just as we see a Grallatorial bird modified into an Ostrich, and a Diver into a Penguin. Now we are told that Mauritius, an island forty miles in length and about one hundred miles from the nearest land, was, when discovered, clothed with dense forests of palms and various other trees. A bird adapted to feed on the fruits produced by these forests would, in that equable climate, have no occasion to migrate to distant lands; it would revel in the perpetual luxuriance of tropical vegetation, and would have but little need of locomotion. Why then should it have the means of flying? Such a bird might wander from tree to tree, tearing with its powerful beak the fruits which strewed the ground, and digesting their stony kernels with its powerful gizzard, enjoying tranquillity and abundance, until the arrival of Man destroyed the balance of Animal Life, and put a term to its existence. Such, in my opinion, was the Dodo, a colossal, brevipennate, frugivorous PIGEON.[1]

The first idea of referring the Dodo to the neighbourhood of the Pigeons, originated with Professor J. T. Reinhardt of Copenhagen, the discoverer of the cranium in the Gottorf Museum. When I was at Copenhagen in 1845, Professor Reinhardt was then absent on a

[1] Mr. E. Blyth, in an excellent treatise on the *Columbidæ* (Journ. As. Soc. Beng. vol. xiv. p. 858, and Ann. Nat. Hist. vol. xix. p. 99), speaking of the *Gourinæ* or Ground Pigeons, says: "Some much resemble Partridges in their mode of life; * * * * other genera are completely sylvan in their abode, *feeding on the ground, more especially on fallen fruits and berries.* Such are the magnificent *Gouras* of the Moluccas and New Guinea, * * * * and the elegant hackled Ground Pigeons (*Calœnas*), one of which abounds in the forests of the Malay Peninsula, and in the Nicobar, Andaman, and Cocos Isles."

voyage round the world, but I was orally informed that he considered the Dodo to be intermediate between the Pigeons and the Gallinaceous birds. On subsequently examining the remains which we possess in Britain, I soon saw reasons for classing this bird even nearer to the Pigeons than I then understood it to be placed by Professor Reinhardt. This gentleman, however, has lately visited London on his return from his distant voyage, and has informed me that, before he left Denmark in 1845, he had pointed out, in his letters to several Swedish and Danish zoologists, "the striking affinity which exists between this extinct bird and the Pigeons, especially the Trerons."

I will now briefly notice the points of agreement in the structure of the Dodo, and in that of the Pigeons, which serve to substantiate the above hypothesis.

A. *External characters.*—1. The whole group of Pigeons are remarkable for having the corneous portion of the beak very short, the basal portion long, slender, and covered with a soft naked skin, all which characters exist in the Dodo, but not in the Gallinaceous birds, nor, with the exception of the *Cathartinæ*, in the Raptores. In all birds the basal portion of the mandibles, whether feathered or bare, is divided from the corneous termination by a separating line; but in the Raptores this basal portion, instead of being depressed, soft, and vascular, as in the Dodo and the Pigeons, is prominent and somewhat hard and horny, resembling wax in appearance, whence it has received the name of *cere*. The *Cathartinæ* are the only Raptores which have a soft cere, and in this very superficial character they may certainly be said to resemble the Dodo.

2. In some species of *Treron*, in *Geophaps*, *Macropygia*, and other Columbine genera, the eyes are surrounded by a naked skin, which, if extended over the face, so as to join the bare basal portion of the beak, would produce the appearance which we see in the *Didus*. In those rare genera, *Verrulia* and *Didunculus* (see plate VII.), this junction of the ocular and rostral areæ actually takes place, and a little more expansion of this naked surface over the forehead would transform those birds into miniature Dodos.

3. In the two strongest beaked genera of Pigeons, *Treron* and *Didunculus*, the corneous portion of the beak is strongly uncinate and compressed, while the tip of the lower mandible curves upwards, and is overhung by the upper one. A comparison of plates V. and VII. will show how precisely this conformation is repeated in *Didus*.

4. In *Treron* and in *Didus* the nostril is placed about the middle of the beak, close to the base of the corneous portion, and near the lower margin. This *forward* and *low* position of the nostril occurs more or less in other genera of Pigeons, but in no other family of birds, that I know of. Some of the Vultures have this orifice equally forward, but none so low down as *Treron* or *Didus*. (See plate VII. fig. 3). Nor can any stress be laid on the supposed absence of an incumbent scale in the Dodo ("sans écaille supérieure"), referred to by M. De Blainville as a Vulturine character. The only meaning which we can attach to the phrase, "nostrils furnished with an incumbent scale," often met with in Bird-books, is that the nostrils enter the beak *obliquely*, so that their upper margin overhangs the lower. Now this is, in fact, the case in the Dodo, whose nostrils are remarkably oblique, and are overhung above by a soft, tumid skin, agreeing herein with the Pigeons, and differing from the Raptores.

5. We find in the Pigeons, even to a greater degree than in *Didus*, that sudden sinking from the forehead to the beak, and the rapid narrowing of the beak in front of the orbits, which Professor

Owen points out as a distinction between the latter bird and the Vultures. This is the result of the prolongation of the beak, and the approach to parallelism in its opposite surfaces; and the only other birds which exhibit this conformation, are certain *Grallatores* and *Tenuirostres*, neither of which groups have any reference to the present question.

6. The apparent width of gape is one of the characters referred to by De Blainville in proof of the Vulturine relations of the Dodo. But the fact is, that the rictus of the Dodo is by no means so wide as in the Raptorial birds, and is proportionably no wider than in the Pigeons. On examining the original specimen, the angle of the mouth is seen to terminate three quarters of an inch in front of the eye. From this point a remarkable cutaneous ridge, which seems peculiar to this bird, extends backwards and downwards beneath the eye, and gives the appearance of a very capacious mouth. (See plate V).

7. The tarsi of the Dodo are only partially covered with transverse scuta, the upper portion being clothed with small scales. This structure is used by De Blainville as an argument for its affinity to the Vultures, in which the tarsi and greater part of the toes are wholly squamose. But although in the majority of Pigeons the tarsi are covered anteriorly with transverse scuta, yet it is interesting to find that in two genera, *Starnænas* and *Goura*, whose habits are almost wholly terrestrial, we find the tarsi clothed with small scales, not unlike those in the Dodo.

8. The absence of metatarsal spines which has been adduced as an objection to the supposed Gallinaceous affinities of the Dodo, prevails equally throughout the *Columbidæ*.

9. The short robust tarsi, and broad expansion of the lower surface of the toes in the Dodo (see Pl. VI) are much more conspicuous in the Pigeons, especially in the group *Treroninæ* (including *Carpophaga*), than in the Vultures. I know no other group in which the toes are similarly expanded, except the Hornbills (*Bucerotidæ*), and these assuredly have no affinity to the Dodo. The design of this structure is probably to give the bird a firmer footing, and to compensate for the shortness, or insufficient lateral movement of the toes.

10. A general character of *perching* birds consists in the hind toe being articulated so low down, that its inferior surface forms a continuous plane with the sole of the foot; whereas in those orders which are essentially ambulatory, such as the *Rasores* and *Grallatores*, the hind toe is more or less raised above the level of the other toes. But in the Pigeons, whose habits are essentially arboreal, the former structure is constant, even in the strictly terrestrial genera, and in the case of the Dodo, although it must have been exclusively confined to the ground, Nature still adheres to the Columbine position of the hind toe. An analogous persistence of type is seen in the *Ground Parrots* and *Ground Cuckoos*, in which the reversed position of the outer toe, an essentially *scansorial* structure, is maintained in spite of the discordance of habits.

11. On comparing the relative lengths of the anterior toes in the different genera of Pigeons, with reference to their peculiarities of habit, we find that in the exclusively arboreal genera (such as *Treron*, *Carpophaga*, *Ptilonopus*, &c.), the inner toe is *shorter* than the outer; in the more terrestrial genera (as *Phaps*, *Geophaps*, &c.), it is *longer* than the outer; while in those genera which combine both modes of life, (as *Columba*, *Turtur*, *Geopelia*, &c.), these digits are nearly equal. Conformably with this, we find that in the Dodo, the most terrestrial of all Pigeons, the inner toe is considerably *longer* than the outer. Now although the head of the Dodo agrees most nearly with that of the *Trerons*, from which I infer that it fed, like those birds, on tropical fruits, yet as the *Trerons* are exclusively arboreal birds, it is interesting to observe that the structure of its foot approaches rather to that of the Ground Pigeons.

12. The Dodo, like the Pigeons, is destitute of any membrane between the toes; whereas all the Vultures, as well as the Gallinaceous birds, are characterized by a short interdigital web.

13. The short, strong, blunt claws of *Didus* do not indicate any Raptorial propensities, but are merely such as we find in most ground birds, as in the terrestrial genera of Pigeons, as well as in the *Gallinaceæ*.

B. *Internal Characters.*—14. An argument which has often been used to prove that the Dodo was a Vulture, or, at least, that it was carnivorous, is the toughness and supposed bad taste of its flesh. Tough it undoubtedly was, and so are all large birds. The toughest bird I ever *tried* to eat, was a wild Swan, yet no one would argue from this that Swans are not allied to Geese and Ducks. Even common Wood-Pigeons are by no means remarkably tender. And the alleged bad taste of the Dodo is a pure invention of the moderns, founded on the statement in Van Neck's Voyage, (see p. 9, *supra,*) that the Dutchmen became disgusted with these birds, and called them *Walckvögel*. But this disgust is expressly attributed, first, to their toughness (accompanied, however, with the admission that the breasts and *stomachs* [imagine the taste of a *Vulture's stomach*!] were "saporis jucundi et masticationis facilis"); and, secondly, because they found an abundance of Turtle-Doves which they liked better. And no wonder; Dutch sailors now-a-days, if supplied *ad libitum* with Turtle-Doves and Wood-Pigeons, would doubtless devour the former, and call the latter *Walckvögel*. The voyagers who followed Van Neck seem to have been less dainty, for they both feasted on fresh Dodos, and stored them among their salt provisions (*supra*, pp. 15, 17). It is therefore clear that the little which we ever shall know concerning the flavour of Dodo-meat affords no objection to the Columbine hypothesis.

15. It appears from the paintings of the Dodo, that this bird must have had a very large œsophagal dilatation or crop. This is a structure which occurs in many different orders, its object being, in some cases (as in granivorous birds), to macerate the food before it passes into the stomach; in others (as in the Raptores), to enable the bird to swallow large quantities of food at distant intervals. The crop of the Dodo, therefore, does not prove much as to its affinities, but as there are no birds in which the crop is more developed than in the Pigeons, the figures of the Dodo are quite consistent with its supposed relation to that family.

16. We do not know much as to the degree of muscularity of the Dodo's gizzard. If by the "stomach," (*venter, ventriculus, estomach, maag,*) which the old voyagers found tender and palatable, the *gizzard* is intended, it would certainly imply a small degree of muscular rigidity. This, however, can hardly have been the case, for we are assured by numerous witnesses (*supra*, pp. 12, 15, 17, 20, 22,) that the Dodo had stones in its gizzard; a character which is always accompanied by a very muscular condition of that organ. Be this as it may, we know that stones are only swallowed by frugivorous birds, which require them to triturate their food, and are never found in the gizzards of the Raptores.

17. We are told by Cauche that the Dodo laid only one egg, and the analogous case of the Solitaire (mentioned hereafter), confirms his statement. Now the Gallinaceous birds are generally remarkable for laying a large number of eggs. Raptorial birds, indeed, lay but few, yet no species of that order (as far as I am aware) lays a single egg, like the Dodo. But in the Pigeons we find that a very small number of eggs (commonly *two*) are the prevailing rule, while in certain genera (*Carpophaga* and *Ectopistes*, see Blyth in Journ. As. Soc. Beng. vol. xiv. p. 855), a single egg is produced, as in *Didus*.

There yet remain several osteological peculiarities in the Dodo which are strongly demonstrative of its affinity to the *Columbidæ*, and of its remoteness from the Raptores. But as these will form the subject of the second part of this work, and will there be treated in full detail by Dr. Melville, I will only briefly enumerate the more important ones. These are:—18, the absence or non-development of the vomer, and of the bony septum of the nostrils; 19, the long narrow nasal fissures; 20, the form of the posterior facet of the lower jaw; 21, the oblique direction of the zygomatic bone; 22, the peculiar form of the palatine bones; 23, the mesial occipital foramen above the *foramen magnum*, (peculiar, it would seem, to the Pigeons and the Dodo); 24, the breadth and peculiar twist of the metatarsal of the hind toe (see Plate XI.); 25, the oval transverse section of the tarso-metarsal; 26, the peculiar form of the upper extremity of the tarso-metatarsal, including the arrangement of the calcaneal processes, and of the canals for the passage of the flexor tendons; and 27, the fact (peculiar to the Pigeons and the Dodo) that these canals pass on the *outside* of the posterior ridge of the tarsus, and not on the *inside*, as in Gallinaceous birds.

Such are the principal points of agreement between the Dodo and the Pigeon family, and it will be admitted that they are neither few nor trivial. There are, however, two or three points of diversity which it is only fair to mention.

1. I need only allude as a matter of form to the non-development of wings, as it is admitted on all hands that this character distinguishes the Dodo from all other birds with which it can be legitimately compared, and is as much opposed to the normal structure of the Rapacious birds, as to that of the *Columbidæ*.

2. The small size of the cranium in proportion to the beak distinguishes the Dodo no less from the Pigeons than from the Vultures. This peculiarity results from the small relative dimensions of the brain and eyes. It is a general law that animals of great magnitude (the Elephant and Whale, for instance,) do not require those important organs to be enlarged in the same proportion as the parts destined for locomotion, and the nutritive functions.[1] We need not, therefore, wonder that so colossal a bird as the Dodo should differ in this respect from other members of that family to which it is nearest allied.

3. The Dodo is, as Professor Owen remarks, "peculiar among birds for the equality of length of the metatarsus and proximal phalanx of the hind toe," while in most birds this phalanx is considerably longer than the metatarsal which supports it. The fact is, however, that no argument as to the general affinities of a doubtful ornithic genus can be drawn from the relative proportions of the tarso-metatarsal, the posterior metatarsal, and the proximal phalanx; these proportions varying in each genus according as its habits are more or less cursorial, ambulatory, or insessorial. A glance at Plate XI., where the forms of these bones in five different genera of Pigeons are exhibited, will substantiate this remark.

4. And, lastly, the nostril of the Dodo, although agreeing in *position* with that of *Treron*, is of a

[1] This law is probably based on the distinction between ponderable and imponderable substances. The bones and muscles of an animal are mechanical structures, the size of which bears an exact arithmetical relation to the masses which they are required to move; but the eye and the brain have to deal with light and the nervous fluid—imponderable agents, to which the ordinary laws of mechanics do not apply.

different *form,* being slightly oblique upwards and backwards, while that of *Treron* is more horizontal. This difference, however, is not greater than what prevails in the nostrils of other genera of pigeons.

It appears then, that the only points in which the Dodo can be said to differ materially from the type of the Pigeons, are few in number, and are not such as to make any approximation to the Raptorial form; and I think it will be granted that the numerous and important characters which have been above noticed, will warrant us in regarding the genus *Didus* as a very aberrant member of the family *Columbidæ*.

Postscript 1.—At pp. 25, 33, *supra,* I have inadvertently spoken of "the Gottorf Museum at Copenhagen." At the time when Olearius published his catalogue, this collection was not at Copenhagen, but at Gottorf, the seat of the Dukes of Schleswig; whence it was removed by Frederic IV., about 1720, to Copenhagen, and was incorporated with the Royal "Kunstkammer" in that metropolis.

2. It has been suggested to me that *translations* of the Latin, French, Dutch, and German passages, extracted above (pp. 9–25), would be acceptable to many readers, and these are therefore given in the Appendix.

CHAPTER II.

The Brevipennate Bird of Rodriguez, the SOLITAIRE.

(*Pezophaps solitaria*, nobis;—*Didus solitarius* of Gmelin.)

Evidence of Leguat; of Herbert—Bones sent to the Paris Museum; to the Andersonian Museum at Glasgow; to the Zoological Society of London—Affinities of the Solitaire.

I NOW proceed to notice another bird of equally remarkable structure to the Dodo, and the evidence, both historical and osteological, of whose existence, though less abundant, is equally positive. The Island of Rodriguez, which is about fifteen miles long by six broad, and situated about three hundred miles to the east of Mauritius, gave birth to an apterous bird called the *Solitaire*,[1] which seems to have been an homologous representative of the Dodo in the last-mentioned island.[2] Rodriguez appears to have remained in a desert and uninhabited condition until 1691, when a party of French Protestant refugees settled upon the island, and remained there for two years. Their commander, François Leguat, a man of intelligence and education, has left a highly interesting account of their adventures, and of the various productions of the island. The chief portion of his work which concerns us at present I will extract in the French original, accompanied by an old translation.

[1] The name *Solitaire* had originally been given to an allied, though doubtless distinct, bird in Bourbon, of which we shall speak presently. Leguat (who never visited Bourbon) probably supposed the Rodriguez bird to be the same species, and therefore gave it the name which other voyagers had imposed on the Bourbon bird. But as Leguat's bird is the type of the "*Didus solitarius*" of systematists, I prefer retaining for it *par excellence* the name of *Solitaire*.

[2] Representation in Zoology is of two kinds, *analogous* and *homologous*. Analogous representation is where a group or species in *one part* of the organic creation performs a similar office, and is, *quoad hoc*, similarly organized, to a group or species in *another part*: e.g., the *Cetacea* among Mammals represent *by analogy* the Fish among *Vertebrata*. This kind of representation exists irrespectively of time and space. Homologous representation is where two groups or species in the *same part* of the organic creation perform a similar office in different geographical regions, or at different times. Thus the Elephants of India and of Africa represent each other *by homology* in space, as the Mammoth and modern Elephants do in time. See Philosophical Magazine, Ser. 3. vol. xxviii. p. 354.

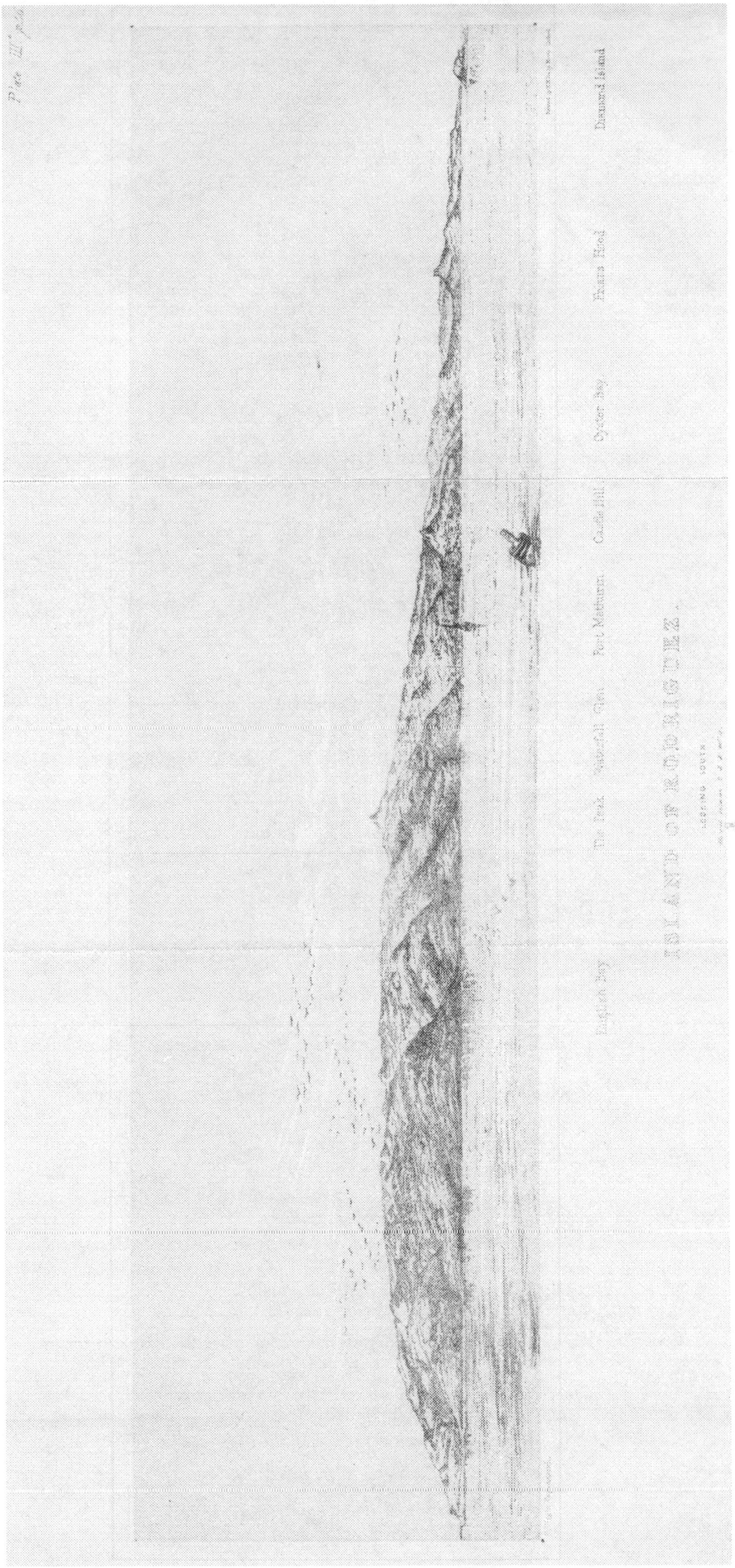

The material originally positioned here is too large for reproduction in this reissue. A PDF can be downloaded from the web address given on page iv of this book, by clicking on 'Resources Available'.

Voyages et Avantures de François Leguat. 2 vols. 12 mo., London, 1720.(2nd ed.)

"De tous les oiseaux de cet isle l'espèce la plus remarquable est celle à laquelle on a donné le nom de *Solitaires,* parce qu'on les voit rarement en troupes, quoique il y en a beaucoup. Les mâles ont le plumage ordinairement grisâtre et brun, les pieds de coq d'Inde, et le bec aussi, mais un peu plus crochu. Ils n'ont presque point de queue, et leur derrière couvert de plumes est arrondi comme une croupe de cheval. Ils sont plus haut montés que les coqs d'Inde, et ont le cou droit, un peu plus long à proportion que ne l'a cet oiseau quand il leve la tête. L'œil noir et vif, et la tête sans crête ni houpe. Ils ne volent point, leurs ailes sont trop petites pour soutenir le poids de leurs corps. Ils ne s'en servent que pour se battre et pour *faire le moulinet* quand ils veulent s'appeller l'un l'autre. Ils font avec vîtesse 20 ou 30 pirouettes tout de suite du même côté, pendant l'espace de 4 ou 5 minutes: le mouvement de leurs ailes fait alors un bruit qui approche fort de celui d'une *Crécerelle,* et on l'entend de plus de 200 pas. L'os de l'ailleron grossit à l'extremité et forme sous la plume une petite masse ronde comme une balle de mousquet, cela et le bec sont la principale defense de cet oiseau. On a bien de peine à les attrapper dans les bois, mais comme on court plus vîte qu'eux, dans les lieux degagés, il n'est pas fort difficile d'en prendre. Quelquefois même on en approche fort aisement. Depuis le mois de Mars jusqu' au mois de Septembre ils sont extraordinairement gras, & le goût en est excellent sur tout quand ils sont jeunes. On trouve des mâles qui pesent jusqu' à 45 livres.

"La femelle est d'une beauté admirable, il y en a de blondes & de brunes; j'appelle blond, une couleur de cheveux blonds. Elles ont une espèce de bandeau comme un bandeau de veuves au haut du bec, qui est de couleur tannée. Une plume ne passe pas l'autre sur tout leur corps, parce qu'elles ont un grand besoin de les ajuster, & de se polir avec le bec. Les plumes qui accompagnent les cuisses sont arron-

A new Voyage to the East Indies by Francis Leguat and his Companions. 12 mo. London, 1708.

"Of all the Birds in the Island the most remarkable is that which goes by the name of the *Solitary,* because it is very seldom seen in Company, tho' there are abundance of them. The Feathers of the Males are of a brown grey Colour: the Feet and Beak are like a Turkey's, but a little more crooked. They have scarce any Tail, but their Hind-part covered with Feathers is roundish, like the Crupper of a Horse; they are taller than Turkeys. Their Neck is straight, and a little longer in proportion than a Turkey's when it lifts up his Head. Its Eye is black and lively, and its Head without Comb or Cop. They never fly, their Wings are too little to support the weight of their Bodies; they serve only to beat themselves, and flutter when they call one another. They will whirl about for twenty or thirty times together on the same side, during the space of four or five minutes. The motion of their Wings makes then a noise very like that of a Rattle; and one may hear it two hundred Paces off. The Bone of their Wing grows greater towards the Extremity, and forms a little round Mass under the Feathers, as big as a Musket Ball. That and its Beak are the chief Defence of this Bird. 'Tis very hard to catch it in the Woods, but easie in open Places, because we run faster than they, and sometimes we approach them without much Trouble. From *March* to *September* they are extremely fat, and tast admirably well, especially while they are young, some of the Males weigh forty-five Pounds.

"The Femals are wonderfully beautiful, some fair, some brown; I call them fair, because they are of the colour of fair Hair. They have a sort of Peak, like a Widow's upon their Breasts [*lege* Beaks], which is of a dun colour. No one Feather is straggling from the other all over their Bodies, they being very careful to adjust themselves, and make them all even with their Beaks. The Feathers on their Thighs are round like shells at the end, and being there very thick, have an agreeable effect. They have two

dies par le bout en coquilles, et comme elles sont fort épaisses en cet endroit-là, cela produit un agréable effet. Elles ont deux élévations sur le jabot, d'un plumage plus blanc que le reste, & qui represente merveilleusement un beau sein de femme. Elles marchent avec tant de fierté et de bonne grace tout ensemble, qu'on ne peut s'empêcher de les admirer & de les aimer, de sorte que souvent leur bonne mine leur a sauvé la vie."—p. 98.

Risings on their *Craws*, and the Feathers are whiter there than the rest, which livelily represents the fine neck of a Beautiful Woman. They walk with so much Stateliness and good Grace, that one cannot help admiring and loving them; by which means their fine Mein often saves their Lives."—p. 71.

The author then proceeds as follows :—

"Tho' these Birds will sometimes very familiarly come up near enough to one, when we do not run after them, yet they will never grow Tame. As soon as they are caught they shed Tears without Crying, and refuse all manner of Sustenance till they die.

"We find in the Gizards of both Male and Female, a brown Stone, of the bigness of a Hen's Egg, 'tis somewhat rough, flat on one side and round on the other, heavy and hard. We believe this Stone was there when they were hatched, for let them be never so young, you meet with it always. They have never but one of 'em, and besides, the Passage from the Craw to the Gizard is so narrow, that a like Mass of half the Bigness cou'd not pass. It serv'd to whet our Knives better than any other Stone whatsoever. When these Birds build their Nests, they choose a clean Place, gather together some Palm-Leaves for that purpose, and heap them up a foot and a half high from the Ground, on which they sit. They never lay but one Egg, which is much bigger than that of a Goose. The Male and Female both cover it in their turns, and the young is not hatch'd till at seven Weeks' end: All the while they are sitting upon it, or are bringing up their young one, which is not able to provide for itself in several Months, they will not suffer any other Bird of their Species to come within two hundred Yards round of the Place; But what is very singular, is, the Males will never drive away the Females, only when he perceives one he makes a noise with his Wings to call the Female, and she drives the unwelcome Stranger away, not leaving it till 'tis without her Bounds. The Female do's the same as to the Males, whom she leaves to the Male, and he drives them away. We have observ'd this several Times, and I affirm it to be true.

"The Combats between them on this occasion last sometimes pretty long, because the Stranger only turns about, and do's not fly directly from the Nest. However, the others do not forsake it till they have quite driven it out of their Limits. After these Birds have rais'd their young One, and left it to itself, they are always together, which the other Birds are not, and tho' they happen to mingle with other Birds of the same Species, these two Companions never disunite. We have often remark'd, that some Days after the young one leaves the Nest, a Company of thirty or forty brings another young one to it, and the new fledg'd Bird, with its Father and Mother joyning with the Band, march to some bye Place. We frequently follow'd them, and found that afterwards the old ones went each their way alone, or in Couples, and left the two young ones together, which we call'd a *Marriage*.

"This Particularity has something in it which looks a little Fabulous, nevertheless, what I say is sincere Truth, and what I have more than once observ'd with Care and Pleasure."

This description is accompanied by a figure, which at once shews that the Solitaire was a very different bird from the *Dodo*; and its accuracy is attested by the fact that in a

Plate IV. p. 48.

Fac-simile of the Frontispiece of Leguat's Voyage.

landscape (see plate IV.) and two maps which accompany the work, no less than twenty-eight small figures of Solitaires are introduced, all of which very closely correspond with the enlarged representation here exhibited.

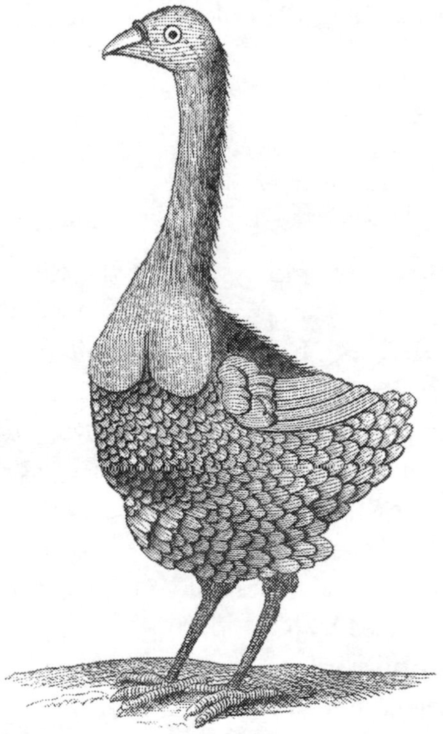

Besides the above lengthened description, Leguat alludes to these birds in several other passages. One of these is very important, as supplying the only testimony extant as to the *food* of any member of the sub-family *Didinæ*.

"The Plantane is a sort of Palm-tree. . . . . . The dates of the Plantane are bigger than those of the Palm-tree. Having abundance of better things to feed on, Fish and Flesh, Fruits, &c., we left the dates for the Turtles and other birds, particularly the *Solitaries,* of which we shall hereafter make mention." pp. 60, 61.

The statement that the Solitaire lays but one egg, and that its nest is a heap of palm-leaves, is very interesting, as Cauche makes a similar assertion regarding the Dodo (*supra*, p. 22). Leguat repeats his statement in another place. Speaking of Sea-Fowl, he says :—

"They lay three times a year, and but one egg at a time, like the *Solitaries*: which is the more remarkable for that if I am not mistaken,[1] we have no example of anything like it among our European Birds." p. 80.

[1] He *was* mistaken, however, for the European Petrels, the Gannet, and most of the *Alcidæ* lay only a single egg.

One more allusion to Solitaires occurs in a sentimental and rather long-winded address, which Leguat makes to the island of Rodriguez on taking his final departure:—

"My Mouth confesses from the abundance of my Heart
That my Soul is touched with Sorrow,
Now I am about to leave thy wholesom Air,
Thy good Palm Wine, thy excellent Melons,
Thy Solitaries, thy Lamentines,
Thy Hills always verdant,
The clear Water of thy Rivers,
Thy fruitful and smiling Sun,
And all thy innocent and rare Delights," &c., &c.—p. 116.

Our only authentic historical evidence respecting the Solitaire is at present confined to Leguat's very circumstantial, though unsupported, testimony. One small item of evidence may indeed be gleaned from Herbert, who sailed past Rodriguez in 1627, but without landing on it, and remarks in his Travels, edition of 1638, p. 341:—

"*Digarroys* [i. e. Rodriguez] . . . . . an ile so desolate; desolate, I mean, in humane inhabitants; other things 'tis uberous in, as wood (choyce and store), Tortoises, Dodos, and other Fowle rare and serviceable." And again, p. 347, speaking of Mauritius:—"Here, and in *Dygarrois* (and nowhere else that ever I could see or heare of) is generated the Dodo," &c. This shows that the existence of an apterous bird in Rodriguez was known in his time, though it was erroneously identified with the Dodo.

Though Rodriguez is a British colony, yet scarcely any information has been published respecting it beyond what Leguat has given us. The island is, however, inhabited by a few colonists, one of whom assured Mr. Telfair that no bird of the kind was now known there (Proc. Z. S. part 1. p. 31). The same negative result was obtained by Edward Higgin, Esq., of Liverpool, who recently suffered shipwreck on this island, and resided there for two months. This gentleman has obligingly favoured me with some MS. notes on Leguat's book, together with other information, which fully establishes the general accuracy of Leguat, though some allowance must be made for that author not having been a naturalist, and for his work having probably been in part written from memory. To Mr. Higgin I am also indebted for the annexed graphic sketch of the scenery of Rodriguez. From the map which Leguat has given of the island, it is evident that the Port of Mathurin, here exhibited, was the site of his settlement, of which we have a view in plate IV.

We cannot, therefore, now hope to procure any living Solitaires, though it would no doubt be perfectly practicable to obtain every part of the skeleton of this bird from the caverns or alluvial deposits of Rodriguez.

If we had no other data than the description and figure of Leguat, we might perhaps refer the Solitaire to the *Struthionidæ* rather than to the Dodo. The legs and neck appear to have been longer, the beak shorter, and the wings, though useless for flight, somewhat more developed than in *Didus*. The short, arched beak, and the defensive structure of the wings, remind us of the Cassowary, rather than of the Dodo. But as we now possess some actual osteological evidences as to its characters, we are enabled to pronounce positively that this bird was closely allied to *Didus*, and was decidedly not Struthious.

As long ago as 1789 certain bones, encrusted with stalagmite, and supposed to belong to the Dodo, were found in a cave in the Island of Rodriguez, by a M. Labistour, whose son-in-law, M. Roquefeuille gave them, about 1830, to the late M. J. Desjardins, Secretary to the "*Société d'Histoire Naturelle de l'Ile Maurice.*" The latter gentleman sent them to Cuvier at Paris, who by some unaccountable confusion of time, place, and circumstance, stated them to have been recently found, under a bed of lava, and in Mauritius. These errors were corrected by M. Desjardins, in the *Analyse des Travaux de la Soc. d'Hist. Nat. de l'Ile Maurice, 2de Année.*[1] (See also Proceedings of Committee of Zoological Society, part 2, p. 111).

It was probably the interest excited by these bones, that induced the late Mr. Telfair in 1831 to apply for further information to Col. Dawkins and to M. Eudes, then resident at Rodriguez. The results of his enquiries are thus recorded in the Proceedings of the Zoological Society, part. 1, p. 31.

"Col. Dawkins, in a recent visit to Rodriguez, conversed with every person whom he met respecting the Dodo, and became convinced that the bird does not exist there. The general statement was that no bird is to be found there, except the Guinea-fowl and Parrot. From one person, however, he learned the existence of another bird, which was called *Oiseau-bœuf*, a name derived from its voice, which resembles that of a Cow. From the description given of it by his informant, Col. Dawkins at first believed that this bird was really the Dodo; but on obtaining a specimen of it, it proved to be a Gannet. It is found only in the most secluded parts of the island.

"Col. Dawkins visited the caverns in which bones have been dug up, and dug in several places, but found only small pieces of bone. A beautiful rich soil forms the ground-work of them, which is from six to eight feet deep, and contains no pebbles. No animal of any description inhabits these caves—not even Bats.

"M. Eudes succeeded in digging up in the large cavern various bones, including some of a large kind of bird, which no longer exists in the island; these he forwarded to Mr. Telfair, by whom they were presented to the Society. The only part of the cavern in which they were found was at the entrance, where the darkness begins; the little attention usually paid to this part by visitors, may be the reason why they have not been previously found. Those near the surface were the least injured, and they occur to the depth of three feet, but nowhere in considerable quantity; whence M. Eudes conjectures that the bird was at all times rare, or, at least, uncommon. A bird of so large a size as that indicated by the bones has never been seen by M. Gory, who has resided forty years on the island.

"M. Eudes adds, that the Dutch, who first landed at Rodriguez, left cats there to destroy the rats which annoyed them; these cats have since become very numerous, and prove highly destructive to poultry; and he suggests it as probable that they may have destroyed the large kind of bird to which the bones belong, by devouring the young ones as soon as they were hatched,—a destruction which may have been completed long before the island was inhabited."

[1] I am indebted to Mr. G. C. Cuninghame for sending me, through Sir W. C. Trevelyan, extracts from the archives of the Mauritian Society, detailing the above facts.

Mr. Telfair having thus procured from Rodriguez a collection of bones, presented one portion of them to the Zoological Society of London, and another to the Andersonian Museum at Glasgow.

Mr. G. C. Cuninghame, of Port Louis, Mauritius, having been recently applied to by Sir W. C. Trevelyan, made several enquiries as to the locality above indicated, and gives a somewhat different account :—

> "I learn that the bones removed [in 1831] were found by digging in a place apparently hollowed out by the action of running water under a mass of rock on the side of a narrow chasm or ravine; that the floor of the cavity is of dark coloured earth, sloping sharply down to its mouth, near which, but *now* considerably below the level of the cavity, a small stream runs at present."

In October 1845, Capt. Kelly, of H.M.S. Conway, made, at Mr. Cuninghame's request, a search for the locality thus indicated. He was unsuccessful in finding the precise spot, but examined two caverns, one of which at the base of a cliff, contained numerous and beautiful stalactites; the other, which he was unable fully to explore for want of a ladder, is in a level piece of ground. The floor of both caves, where not covered with stalagmite, is a fine red mould, which I strongly recommend to the attention of those who may hereafter have the happiness of digging for bones in Rodriguez.

The bones which were sent to Paris were exhibited in 1830 by Cuvier to the Academy of Sciences (Ann. des. Sc. Nat. vol. xxi.; Revue Sept. 103, 104, 109, 110; Bull. Sc. Nat. vol. xxii. p. 122; Ed. Journ. Nat. Sc. vol. iii. p. 30), but no detailed account of them has yet been made public. Being anxious to compare them with the remains of the Dodo which we possess at Oxford, I applied to M. de Blainville to permit these bones to be brought to England. He at once gave his consent, and commissioned Professor Milne Edwards to bring them with him to the meeting of the British Association at Oxford in June 1847, an act of liberality which has enabled Dr. Melville and myself to make the desired comparison.

We were further permitted, by the kindness of the Trustees of the Andersonian Museum at Glasgow, to exhibit to the Association the bones from Rodriguez presented to that institution by the late Mr. Telfair. These gentlemen entrusted the relics to Sir W. Jardine, and allowed him not only to diffuse, by means of plaster casts, the information they convey, but to bring with him the bones themselves to the Meeting.

The bones which were sent by Mr. Telfair in 1833 to the Zoological Society, have met with some unfortunate fate. Three or four years ago, Mr. Fraser, the late Curator of that Society, made at my request a diligent search for these specimens, but all his endeavours to find them were fruitless. Among the many treasures which have been presented to the Society during the last twenty years, and which for want of space are still buried in vaults and outhouses, he found the identical box sent by Mr. Telfair; but, alas! the bones of the Solitaire, apterous as it was, had flown away, and the only bones that remained belonged to *Tortoises*! We are again, therefore, obliged to fall back upon historical records in place of ocular evidence. In the Proceedings of the Zoological Society for March 12, 1833, p. 32, we read that "the bones procured [in Rodriguez] for Mr. Telfair were laid on the table. They include, with numerous bones of the extremities of one or more large species of Tortoise, several bones of the hinder extremity of a large bird, and the head of a humerus. With reference to the metatarsal bone of the bird, which was long and strong, Dr. Grant pointed out that it possessed articulating surfaces for four toes, three directed forwards, and one backwards, as in the foot of the Dodo preserved in the British Museum, to which it was also proportioned in its magnitude and form."

In our attempts, therefore, to reconstruct the skeleton of the Solitaire, and to determine its zoological affinities, our only data are the bones which the Curators of the Paris and Glasgow collections have enabled us to bring into juxta-position. The bones of the supposed Solitaire from the Paris Museum are five in number;[1] viz., a femur, a tarso-metatarsal, a humerus, the medial portion of a sternum, and a portion of the cranium. Unfortunately they are all incrusted uniformly over with stalagmite, from $\frac{1}{16}$ to $\frac{1}{20}$ of an inch in thickness, which prevents all examination of the surface of the bones, or any minute description of their structure. They nevertheless supply us with several important elements to guide us in reconstructing the skeleton of this lost bird.

From the uniformity in the appearance and thickness of the incrustation, it appears evident that these bones have all been obtained in one locality, probably in some pool on the floor of a cavern, exposed to the dripping of water containing carbonate of lime. And from the fact that no duplicate bones occur amongst them, and from their apparent agreement in proportionate size, we have a right to assume that they are portions of the skeleton of the same individual. (See Plates XIII. and XIV.)

The Glasgow series of bones are all portions of the hinder extremity, and consist of three femora, a tibia, and two tarso-metatarsal bones. Their appearance, as well as their history, proves them to have been obtained under different circumstances from those last mentioned. They still contain nearly the whole of their animal matter, present a glossy surface, considerable specific gravity, and are neither changed in colour nor incrusted with extraneous matter. They have the appearance of having been obtained from a reddish soil on the floor of some dry cave, where they have been protected from the changes of weather and from the action of mineral waters.

The only bones which are common to the Paris and Glasgow series are the femur (Plate XIV.) and the tarso-metatarsal. (Plate XV.) On comparing these together, they present every indication of specific identity. The tarso-metatarsal at Paris is of the same form and dimensions (allowing for the thickness of the incrusting matter) as the pair at Glasgow. And the Parisian femur, though apparently much larger, owing to the thickness of its stalagmitic coating, is yet reducible to the same dimensions as the largest of the three Andersonian femora. From this, and from the anatomical relations of the bones to each other, it appears certain that these two collections of bones belong to one and the same species of bird. And

[1] There is a sixth bone in the collection, but it belongs, not to the Solitaire, but to a Tortoise.

as we know that they were all brought from the small island of Rodriguez, where no bird now exists to which they can be referred, we have a right to assume that they belong to the extinct species described and figured by Leguat as the Solitaire.

On comparing these bones from Rodriguez with the few remains extant of the Dodo of Mauritius, we see at once that they are not specifically identical. The tarso-metatarsal from Rodriguez is about an inch longer than that of the Dodo, and the proportions of the other bones indicate a more erect and longer legged bird, precisely as the description and figure of the Solitaire given by Leguat would lead us to expect. On the other hand, the peculiar form of the calcaneal processes, the expansion of the distal end of the tarso-metatarsal, the large surface of attachment for the posterior metatarsal, and other characters which distinguish the Dodo, are precisely repeated in the bones before us, showing that the species to which they belong is unquestionably very nearly allied to, though not identical with, the Dodo. And it is important to remark that as far as we can trace the points of agreement between these two extinct birds, they are shared in common with the Pigeons, and exist *in no other known families of birds.*

Unfortunately the cranium of the supposed Solitaire is very imperfect (see Plate XIII.), and the anterior portion is entirely wanting. With such incomplete data, it may, therefore, appear premature to assert the generic distinction of these two birds. Yet from the greater length of the legs, and less development of the beak, as indicated by Leguat, it seems certain that the Dodo and the Solitaire would be classed (according to the present standard of zoological characters) in two distinct genera. I therefore propose to bestow upon the Solitaire the provisional generic name of PEZOPHAPS (from πέζος, pedestrian, and φάψ, a pigeon), in the confidence that future discoveries of the remaining parts of the skeleton will justify this denomination. The Columbine characters of the Solitaire will be fully described by Dr. Melville in the second Part of this work, but I will draw attention in passing, to certain peculiarities recorded by Leguat in his account of the Solitaire, which confirm this view of its affinities. I refer to the feeding on Dates or Plantains, the monogamous habits, the laying only one egg, the building a nest, and the inability of the nestling to provide for itself. Now the first of these characters is incompatible with any supposed Raptorial affinities, and the four last are opposed to the Gallinaceous hypothesis, but the whole of them are consistent with the habits of that anomalous family, the *Columbidæ.*[1] And as we have osteological evidence of the affinity of the Solitaire to the Dodo, we thus obtain a reflected and collateral proof of the Columbine relations of the latter bird.

There is one remarkable character in the skeleton of the Solitaire which seems opposed to the supposition that it belongs to a brevipennate bird. In ordinary birds the power of flying requires great size and strength in the pectoral muscles, and a largely developed keel

[1] Mr. Blyth tells us that the Pigeons of the genus *Carpophaga* "do not in general lay more than one egg, and certain species invariably but one; in which respect they resemble the celebrated Passenger Pigeon of North America (*Ectopistes migratoria*)."—Journ. Asiat. Soc. Bengal. vol. xiv. p. 855.

upon the sternum for their insertion. But in the Ostriches, where the wings are short and feeble, the pectoral muscles are exceedingly small, and the sternum is destitute of a medial keel. Now in the sternum of the Solitaire we find a considerably developed keel, such as would almost indicate volatile powers. (See Plate XIII.) The shortness of the humerus, however, no less than the positive testimony of Leguat, prove that the bird was wholly unable to rise from the ground. The presence of a sternal keel would therefore appear anomalous, were it not for a circumstance mentioned by Leguat, namely, that the bird used its wings for self-defence, and was able to inflict considerable blows with these members, for which end a corresponding strength of the pectoral muscles, and enlargement of the sternal keel would be required. It is, moreover, evident from the figures handed down to us, both of the Dodo and the Solitaire, that the wings of these birds, though too short for flight, were yet considerably more developed both in size and structure, than is the case in the *Struthionidæ*.

Before leaving the Island of Rodriguez I must call attention to the following passage of Leguat:—

"Nos Gelinottes sont grasses, pendant toute l'anneé, & d' un goût très délicat. Elles sont toutes d' un gris clair, n' y ayant que très peu de différence de plumage, entre les deux sexes. Elles cachent si bien leur nids que nous n' en avons pû découvrir, ni par conséquent goûter leurs Oeufs. Elles ont un ourlet rouge autour de l' œil. Et leur bec qui est droit et pointu, est rouge aussi; long d' environ deux pouces. Elles ne se sçauroient guères voler [" they cannot fly"—Eng. ed.], la graisse les rendant trop pésantes. Si on leur présente quelque chose de rouge, cela les irrite si fort qu' elles viennent l' attaquer pour tâcher de l' emporter; si bien que dans l' ardeur du combat on a occasion de les prendre facilement."—p. 103.

The English translation is as follows:—

"Our Wood-Hens are fat all the year round, and of a most delicate taste: Their colour is always of a bright gray, and there's very little difference in the plumage between the two sexes. They hide their nests so well that we cou'd not find 'em out, and consequently did not tast their eggs. They have a red list about their eyes, their beaks are straight and pointed, near two inches long, and red also. They cannot fly, their fat makes 'em too heavy for it. If you offer them anything that's red, they are so angry that they will fly at you to catch it out of your hand, and in the heat of the combat we had an opportunity to take them with ease."—Eng. ed. p. 75.

The name *Gelinotte* would imply a bird allied in appearance to the Grouse of Europe, but the "straight pointed beak, two inches long," seems to place this bird out of the pale of the Gallinaceous order. I cannot help suspecting that we have here an indication of another brevipennate bird, nearly, if not quite, unable to fly, and related, perhaps by *analogy* only, to the *Didinæ*, while its *affinities* may have pointed towards the *Apteryx*. This conjecture derives probability from the unknown Mauritian bird, figured by Van den Broecke, and by Herbert, and described by Cauche (*vide supra*, pp. 19, 21), and which may have been related to the "*Gelinotte*" of Leguat, especially as the latter mentions *Gelinottes* among the birds of Mauritius, as well as of Rodriguez. Cauche, too, records that his "Poules rouges au bec de

Bécasse," were caught with a red rag like Léguat's *Gelinottes*. On the other hand, Mr. Higgins informs me that a species of *Numida*, or Guinea-Fowl, is now abundant in Rodriguez (introduced probably by the early voyagers), and it is therefore possible that Leguat's description may be intended for this bird, although the discrepancies are considerable. The *Gelinotte* question is therefore open to further investigation, and I would especially recommend it to the attention of the "*Société d'Histoire Naturellé de l'Ile de Maurice.*"

CHAPTER III.

Brevipennate birds of the Isle of Bourbon.

Evidence of Castleton; of Bontekoe; of Carré; of Sieur D. B.; of Billiard; of a British Officer—Indications of a Brevipennate Bird in Madagascar—Review of the whole subject—Analogical case of New Zealand—Conclusion.

THE volcanic island of Bourbon, which lies about one hundred miles to the S.W. of Mauritius, is proved by indubitable evidence to have been inhabited by two species of birds, whose inability to fly, and their consequent rapid extinction, brings them into the same category with the Dodo of Mauritius and the Solitaire of Rodriguez. It will be remembered that Bourbon was discovered between 1502 and 1545 by Mascaregnas, a Portuguese, who called the island by his own name, but seems to have left us no other record of his visit.

1. The earliest notice which concerns us is by Captain Castleton, who visited Bourbon in 1613. In the account of his voyage, written by J. Tatton, one of his officers, we read:—

> "There is store of Land-fowl, both small and great, plentie of Doves, great Parrats, and such like; and a great fowl of the bigness of a Turkie, very fat, and so short winged that they cannot flie, beeing white, and in a manner tame; and so are all other fowles, as having not been troubled nor feared with shot. Our men did beate them down with sticks and stones. Ten men may take fowle enough to serve forty men a day." (Purchas, ed. 1625. vol. i. p. 331. This narrative was also published separately in 1690, and is included in Prevost's Histoire Générale des Voyages, vol. ii. p. 120; in Harris's Voyages, vol. i. p. 115; and in Grant's Mauritius, p. 164.)

2. In 1618, Bontekoe, a Dutch voyager, spent twenty-one days in Bourbon, which he describes as abounding with Geese, Parrots, Pigeons, and other game, and adds, "there were also *Dod-eersen*, which have small wings, and so far from being able to fly, they were so fat that they could scarcely walk, and when they tried to run, they dragged their under side along the ground." The original words, contained in the Journael ofte gendenckwaerdige Beschryvinge van de Oost-Indische Reyse van Willem Ysbrantz Bontekoe van Hoorn, 4to. Rotterdam, 1674, are as follows:—

> "Daer waren oock eenige dod-eersen, die kleyne vleugels hadden, maer konden niet vliegen, waren soo vet datse qualijck gaen konden, want als sie liepen, sleepte haer de neers langhs de aerde."—p. 7.[1]

[1] Bontekoe's Voyage was published in Dutch at Haerlem in 1646, at Rotterdam in 1647, at Utrecht in 1649 and 1651, and at Amsterdam in 1648, 1650, and 1656. A French translation will be found in Thevenot's Relations de divers Voyages Curieux, Paris, 1663, vol. i., and a German one in Hulsius's "Vier und zwanzigste Schiffart," &c. 4to. Franckfort, 1648. p. 7.

Bontekoe appears to have considered these birds identical with the Dodos of Mauritius, and the slowness of pace and shortness of leg, which his description implies, hardly agree with what we know of these Bourbon birds. But as we have no other proof of the existence of the Dodo in Bourbon, and as Bontekoe's account must have been written from memory (for his ship was afterwards blown up, and he was the sole survivor), we must not look for scientific accuracy in his statement. The probability is, that when he in after years compiled the narrative of his perilous adventures, having a recollection of a large brevipennate bird in Bourbon, whose tameness rendered it an easy prey to his sailors, he concluded it to be the Dodo, and adopted the name and descriptions of that bird which had been given by previous navigators.

3. We have next to notice the narrative of a Frenchman, named Carré, who visited Bourbon in 1668, and relates as follows:—

"J'ay vû dans ce lieu une sorte d'oiseau que je n'ay point trouvé ailleurs: c'est celuy que les habitans ont nommé l'Oiseau Solitaire, parce qu'effectivement il aime la solitude, et ne se plait que dans les endroits les plus écartez; on n'en a jamais vû deux ni plusieurs ensemble; il est toujours seul. Il ne ressembleroit pas mal à un Coq d'Inde, s'il n'avoit point les jambes plus hautes. La beauté de son plumage fait plaisir à voir. C'est une couleur changeante qui tire sur le jaune. La chair en est exquise: elle fait un des meilleurs mets de ce païs-là, et pourroit faire les délices de nos tables. Nous voulumes garder deux de ces oiseaux pour les envoyer en France, et les faire présenter à Sa Majesté; mais aussi-tôt qu'ils furent dans le Vaisseau, ils moururent de melancolie, sans vouloir ni boire ni manger."—Voyages des Indes Orientales par M. Carré, 2 vols. 12mo. vol i. p. 12. See also Prevost, Hist. Gén. des Voyages, vol. ix. p. 3.

Translation:—

"I here saw a kind of bird which I have not found elsewhere: it is that which the inhabitants call the *Oiseau Solitaire*, for, in fact, it loves solitude, and only frequents the most secluded places. One never sees two or more of them together; they are always alone. It is not unlike a Turkey, were it not that its legs are longer. The beauty of its plumage is delightful to behold. It is a changeable colour, which verges upon yellow. The flesh is exquisite; it forms one of the best dishes in this country, and might form a dainty at our tables. We wished to keep two of these birds to send to France and present them to His Majesty, but as soon as they were on board ship, they died of melancholy, having refused to eat or drink."

It will be observed that Tatton describes these birds as *white*. Carré's expression, "une couleur changeante qui tire sur le jaune," is rather vague, but seems to imply a pale yellowish or cream-coloured tint, which another author might easily have described as white. At any rate there seems no reasonable doubt that Tatton and Carré both described the same species of bird.

4. In the year after Carré's visit, a French colony was sent from Madagascar to Bourbon under M. de la Haye. One of the party, who calls himself the Sieur D. B., has left an interesting account of the expedition. His journal is contained in a MS., given by Mr. Telfair

to the Zoological Society of London, which I hope will not be allowed to remain much longer unpublished. He not only confirms the accounts given by Tatton, Bontekoe, and Carré, of a brevipennate bird in Bourbon, but gives us a clear proof that a second species of the same group of birds inhabited that island. Speaking of the land-birds of the island, he enumerates,

1. "*Solitaires*: ces oiseaux sont nommés ainsi, parce qu'ils vont toujours seuls. Ils sont gros comme une grosse Oye, et ont le plumage blanc, noir à l'extremité des ailes et de la queue. A la queue il y a des plumes approchantes de celles d'Autruche, ils ont le col long, et le bec fait comme celui des Bécasses, mais plus gros, les jambes et pieds comme poulets d' Inde. Cet oiseau se prend à la course, ne volant que bien peu.

2. "*Oiseaux bleus*, gros comme les *Solitaires*, ont le plumage tout bleu, le bec et les pieds rouges, faits comme pieds de poules, ils ne volent point, mais ils courent extrèmement vîte, tellement qu'un chien a peine d' en attraper à la course; ils sont très bons."[1]

Translation :—

1. "*Solitaires*. These birds are so called because they always go alone. They are the size of a large Goose, and are white, with the tips of the wings and tail black. The tail feathers resemble those of an Ostrich; the neck is long, and the beak is like that of a Woodcock, but larger; the legs and feet like those of Turkeys. This bird has recourse to running, as it flies but very little.

2. "*Oiseaux bleus*, the size of *Solitaires*, have the plumage wholly blue, the beak and feet red, resembling the feet of a hen. They do not fly, but they run extremely fast, so that a dog can hardly overtake them; they are very good eating."

I should have been disposed to refer the "Oiseau bleu" to the genus *Porphyrio*, were we not told that they were the size of the Solitaire, i. e., of a large Goose, that the feet resembled those of a hen, and that they never fly. Moreover, Bory St. Vincent in his list of the Birds of Bourbon (Voy. aux quatre Iles de la Mer d'Afrique, vol. i.), makes no mention of any species of *Porphyrio*.

It is evident from these statements,

1st, That Bourbon was formerly inhabited by a brevipennate bird called the Solitaire, whose white or light yellow plumage, and Woodcock-like beak proves it to have been distinct from the Dodo of Mauritius and from the so-called Solitaire of Rodriguez.

2ndly, The account given by the Sieur D. B. seems to imply that this bird possessed some, though very imperfect, powers of flight; but as Tatton and Bontekoe distinctly assert the contrary, we may presume that this statement of the former author was inaccurate.

And 3rdly, it is clear that a second brevipennate species, the "*Oiseau bleu*" of Sieur D.B., was also a native of Bourbon, though from its speed in running it probably escaped the notice of the earlier voyagers.

5. Of this *Oiseau bleu*, the only other indication which I have met with is in Rees' Cyclopædia, art. "BOURBON," where it is stated that in Bourbon there is "a kind of large

[1] This passage was first published in a paper which I communicated to the Zoological Society, Apl. 23, 1844. (Proc. Z. S. pt. xii. p. 77.)

bat, denominated *l'Oiseau bleu*, which are skinned and eaten as a great delicacy." This is evidently a blunder, as regards the "Oiseau bleu" being a *bat*, but it proves that some author besides the Sieur D. B. has noticed the *Oiseau bleu* of Bourbon, though I have been unable to discover from what work this statement is copied.

6. We have evidence that one, at least, of these apterous species of birds continued to inhabit Bourbon till nearly the middle of the last century. M. Billiard, who resided in that island between 1817 and 1820, and appears to have had access to some of the original archives of the island, tells us that at the time of its first colonization "the woods were filled with birds which were not alarmed at the approach of man. Among these was the *Dodo* or *Solitaire*, which was pursued on foot; they were still to be seen in the time of M. de la Bourdonnaye, who sent a specimen as a curiosity to one of the Directors of the Company." Now M. de la Bourdonnaye was Governor of the Isles of France and Bourbon from 1735 to 1746, so that these singular birds *must* have survived till the former, and *may* have continued till the latter date at least.

7. In Grant's Mauritius, p. 167, is an extract from "Observations on the Isle of Bourbon in 1763, by an Officer of the British Navy," which may possibly indicate that these singular birds survived in that island as recently as the above date:—

> "The plain *des Caffres* is formed by the summits of mountains at a very considerable elevation above the sea. . . . . . On this elevated plain there are small trees, with broom, furze, a kind of wild oat, and fern, which grows to the height of a shrub. There are also some curious birds which never descend to the sea-side, and who are so little accustomed to, or alarmed at, the sight of man, that they suffer themselves to be killed by the stroke of a walking-stick."

Whether the "curious birds" here alluded to, be referable to the brevipennate group or not, does not appear, but it seems certain that in 1801, when Bory St. Vincent made a careful scientific survey of the Island of Bourbon, no such birds were then in existence.[1]

Our evidence respecting the brevipennate birds of Bourbon is at present confined to Historical testimony. No delineations of these creatures appear to be now extant, and their osseous remains have never yet been sought for, and have consequently never yet been found. We cannot therefore at present decide whether these extinct birds were more allied to the Dodo of Mauritius, or to the Ostrich of Africa, though from the descriptions given, the former supposition is most probable. We naturally look to the little-known island of Madagascar as the region most likely to contain birds allied by affinity to those of Bourbon. No recent

[1] The reader must beware of adducing an additional testimony from a passage which that careless compiler, Grant, in his chapter on Bourbon, professes to quote from Du Quesne:—"The Giant *and the Dodo* are large birds of an extraordinary height, which frequent the rivers and lakes, and whose flesh is like that of the Bittern." (Hist. of Mauritius, p. 154.) In Du Quesne's account of Bourbon (drawn up apparently as an *emigrant-trap*) as quoted by Leguat, p. 56 (for I have not been able to find the original), the words are "Les Géans sont de grands oiseaux montés sur des échasses," &c. The words "and the Dodo" are therefore an interpolation of Grant's, nor does the English translator of Leguat mend the matter (p. 41), by rendering *Géans* into *Peacocks*! The fact is, that these *Géans* are evidently (notwithstanding the Stork-like aspect of Leguat's plate at p. 171) *Flamingos.*

travellers have alluded to the existence of any Struthious or brevipennate birds in Madagascar, though from the following passage in Flacourt's *Histoire de la grande Isle Madagascar*, published at Paris in 1658, 4to., it appears that a bird of that family inhabited Madagascar less than two centuries ago. Flacourt tells us that "the *Vouron patra* is a large bird which frequents the region of Ampatres [a province at the south extremity of Madagascar] and lays eggs like the Ostrich. It is a kind of Ostrich; the inhabitants are unable to capture it, and it inhabits the most desert places."

> "Oyseaux qui hantent les bois. *Vouron patra*, c'est un grand oyseau qui hante les *Ampatres* et fait des œufs comme l'Autruche; c'est une espèce d'Autruche; ceux des dits lieux ne le peuvent prendre; il cherche les lieux les plus déserts."—p. 165.

This brief indication may perhaps guide the future explorer of Madagascar to a discovery of great zoological interest.

On a review of the various Historical and Osteological evidences which I have now brought together, it seems sufficiently clear that the three oceanic islands, Mauritius, Rodriguez, and Bourbon, which, though somewhat remote from each other, may be considered as forming one geographical group, were inhabited, until the time of their human colonization, by *at least four* distinct, but probably allied, species of brevipennate birds. This result at once reminds us of the analogous case of the New Zealand group of islands, where the scientific zeal of Messrs. Cotton, Williams, Colenso, Mantell, and others, has brought to light a mine of osteological treasures, from which the consummate sagacity of Prof. Owen has re-constructed two new genera of brevipennate birds. Seven species of *Dinornis* and two of *Palapteryx* have been clearly established and elaborately described by Professor Owen, while in the still surviving genus *Apteryx*, of which Mr. Gould has very recently described a *second* species, we see an almost expiring member of the same zoological group.[1]

The extraordinary success of the naturalists of New Zealand, in procuring from recent alluvial deposits a series of osseous remains which have more than doubled the number of *Struthioid* birds previously known, should encourage the scientific residents in the islands of the Indo-African Sea to make similar researches. I feel confident that if an active naturalist would make a series of excavations in the alluvial deposits, in the beds of streams, and amid the ruins of old habitations in Mauritius, Bourbon, and Rodriguez, he would speedily discover remains of the *Dodo*, the two "*Solitaires*," or the "*Oiseau bleu*." But I would

[1] The recent discovery of the heads of *Dinornis* and *Palapteryx* has proved that these two genera are not so nearly allied as was at first supposed. Professor Owen read a paper on the subject to the Zoological Society, January 11th, 1848, in which he shows that "the beak of *Palapteryx* is decidedly Struthious. The beak and skull of *Dinornis* differ very essentially from any form, either recent or extinct."—(*Athenæum*, no. 1057, p. 116). In a recent communication to the Geological Society, Feb. 2nd, 1848, Dr. Mantell states that the ornithic bones sent by his son from New Zealand are referable to no less than *five* genera.—(*Athenæum*, no. 1061, p. 218).

especially direct the attention to the caves with which those volcanic islands abound. The chief agents in the destruction of the brevipennate birds were probably the run-away negros, who for many years infested the primæval forests of those islands, and inhabited the caverns, where they would doubtless leave the scattered bones of the animals on which they fed. Here, then, may we more especially hope to find the osseous remains of these remarkable animals.

Should any copies of this work find their way to Mauritius or Bourbon, they may perhaps incite the lovers of knowledge in those islands to investigate further the subject which has been diligently, but imperfectly, pursued in this volume. And I shall feel rewarded for the trouble it has cost, if my researches into the history and organization of these birds, aided by the anatomical investigations which Dr. Melville has introduced into the second part of the work, shall have rescued these anomalous creatures from the domain of Fiction, and established their true rank in the Scheme of Creation.

END OF PART I.

Postscript to Part I.

THE foregoing sheets had been printed some time, and the second part of this work had been unavoidably delayed by the great attention which the osteological plates and descriptions required, when I was led to some additional sources of information which demand notice.

The first of these is a rare edition of Bontekoe's Voyage, kindly communicated to me by Dr. Bandinel, the Bodleian Librarian, entitled "Journael van de acht-jarige avontuerlijcke Reyse van Willem Ysbrantsz Bontekoe van Hoorn, gedaen nae Oôst-Indien," published in 4to at Amsterdam, by Gillis Joosten Zaagman. There is no date, but from a narrative introduced at the end, it must be subsequent (probably only by a year or two) to 1646. The narrative is nearly a verbatim version of the other Dutch editions of Bontekoe (noticed at p. 57 *supra*), and the only variation of text which concerns us, is in the statement that the underside of the Dodo dragged along the ground, which is here qualified thus:—"sleepte haer de neers *by na* (i. e. *almost*) langs de Aerde." But what gives a peculiar interest to this volume is, that it contains (alone of all the editions of Bontekoe which I have seen) a figure of the Dodo, which I here present.

This highly ludicrous representation is more like a Fighting-cock than a Dodo, and the black-letter of the Dutch text omits to tell us whether this design was due to the pencil of Bontekoe or his publisher Zaagman, or whether it was copied from some contemporary painting now forgotten. But there can be no doubt that this figure refers to the true Dodo of Mauritius, and not to the "Solitaire" of Bourbon, with which Bontekoe confounded it (see p. 58 *supra*).

We may regret that the rudeness of the original woodcut leaves us in the dark as to the nature of the object on which the Dodo appears about to feed. This figure would pass equally well for a testaceous mollusc, or for an arboreal fruit, so that the problem of the Dodo's food seems as far from a solution as ever.

A notice of Savery's Dodo-picture in the Belvedere at Vienna (see p. 30 *supra*) is given in the Archiv für Naturgeschichte, for 1848, p. 79, by Dr. L. J. Fitzinger, who there states that he has long known this interesting painting, and was on the point of publishing a fac-simile of it, when, hearing that this work was in course of preparation, he courteously resigned his intention, and contented himself with publishing a brief notice of its existence. He states that this picture measures sixteen by twenty-two inches, and represents an ideal landscape with the fore-ground crowded with birds, some on land, and some in the water, all of which are accurately designed.

Five weeks had elapsed since the last sheets of Part I. had gone to press, when, on May 16th, 1848, I received (through the kindness of my friend and former fellow-traveller, Mr. W. J. Hamilton, P.R.G.S.) a pamphlet by Dr. Hamel, entitled "Der Dodo, die Einsiedler, und der erdichtete Nazarvogel." I am thus exact as to dates, in order that the similarity between many of Dr. Hamel's inferences and my own may be attributed, not to plagiarism, but to the Unity which characterizes Truth. This memoir was read before the Petersburg Academy on January 9th, 1846, but has only just been published in the Bulletin Phys.-math. Acad. St. Petersb. vol. vii. no. 5, 6. Dr. Hamel here gives a *resumé* of the historical and pictorial evidences respecting the Dodo and Solitaire, as far as he had ascertained them, but he leaves untouched the question of their affinities, and too often omits to indicate the original sources of his information. As I have already discussed most of the details contained in this treatise, I need only refer to two or three points which had escaped my notice.

The diligent researches of Dr. Hamel appear to have added nothing to the historical evidence which is recorded above. The only work mentioned by him which I had failed to consult is the Journal of Paul van Soldt, for which I had sought in the libraries of Oxford and London without success. This, however, is merely another version of the account of Van der Hagen's Voyage, and does not add to the information respecting it given at p. 17 *supra*.

Dr. H. has judiciously remarked that from an obscurity of expression in the earliest account of Van Neck's Voyage, the Dodo was described by translators and subsequent compilers as having the wings blackish and the tail grey. But we know from the coloured paintings that the whole bird was greyish, and the wings and tail yellowish. (See Plates I., III.) This error was corrected by Matelief (p. 17 *supra*), who stated the plumage to be grey, and by Verhuffen (or rather his officer and journalist Verkens), in whose narrative (p. 18 *supra*) it is added that the wing feathers were yellow.

Dr. Hamel has shewn the probability that the island, or bank, of Nazareth (see p. 21 *supra*) has no more existence than the *Didus nazarenus* to which it gave a name. I must therefore apologize to geographers for having introduced this *vigia* into the chart of the Indo-African Ocean at p. 6, which was copied from Mr. Arrowsmith's map of the world, published in 1842.

The *Géans* of Leguat, which I have referred to Flamingos (p. 60 *supra*), are by Dr. Hamel conjectured to be Struthious birds, which, like the Solitaire, have become extinct since the days of Leguat. On re-perusing Leguat's text, however, it does not appear to me that the discrepancies between his *Géans* and the Flamingo are so great as to justify this conclusion.

After quoting Leguat's account of the Solitaire, Dr. Hamel tells us the following anecdote. The French astronomer Pingré visited Rodriguez in 1761, to observe the famous transit of Venus, which was the occasion of many similar expeditions. To commemorate this circumstance Le Monnier proposed to place the Solitaire among the constellations, but being a better astronomer than ornithologist, he inadvertently gave this honour, not to the Didine bird of Rodriguez, but to the Solitary Thrush of the Philippines (*Monticola eremita*), figured by Brisson, vol. ii. pl. 28. f. 1, instead of copying Leguat's figure as he might have done. (See Mémoires de l'Académie, 1776, p. 562, pl. 17.) It is worth the consideration of astronomers whether the imaginary outline of this constellation might not be so altered as to restore to Leguat's Solitaire the honours which are its due.[1]

In connection with Pingré's visit, Dr. Hamel adds the following judicious suggestion:—"We know the spot in Rodriguez where Leguat and his companions resided for two years. It appears that Pingré also lived there in 1760 and 1761, and conducted his astronomical observations, for he says (Hist. de l'Acad. 1761, p. 108, and Mémoires, p. 415) that the place was called 'Enfoncement de François Leguat.'[2] In Leguat's map the place is accurately indicated where the common kitchen of the settlers stood, and where the great tree grew, under which they used to sit on a bench to take their meals. The tree and bench are introduced in the map. At these two spots it is probable that the bones for a complete skeleton of Leguat's Solitaire might be collected; those of the head and feet on the site of the kitchen, and the sternum and other bones on that of the tree."

I have next to notice a memoir by Professor Owen, just published in the Transactions of the Zoological Society, vol. iii. p. 345, on the remains of *Dinornis, Palapteryx, Notornis,* and *Nestor,* discovered by Mr. W. Mantell in New Zealand. In this paper Professor Owen has availed himself of the recent dissection of the Dodo's head, to carry on the comparison of that bird with the *Dinornis,* which he had commenced (in regard to the leg bones) in 1846. He further remarks: "With respect to the Dodo, the idea entertained by Dr. Reinhardt and by Mr. Gould[3] of its affinity to the *Columbidæ,* was supported by new arguments adduced by Mr. Strickland in his elaborate and interesting communications and lecture before the British Association at Oxford (June, 1847)."

This quotation contains a slight inaccuracy which I must be allowed to correct. In regard to Professor Reinhardt, I have already (at p. 40 *supra*) acknowledged the originality of his idea as to the affinity between the Dodo and the *Columbidæ,* but there is no trace of this idea in any of Mr. Gould's published writings. It is true that in his account of the *Gnathodon,* published March 1st, 1846 (see p. 40 *supra*), Mr. G. was the first to assert its affinity to the Pigeons, and he at the same time incidentally adds that the form of the beak and nostrils "strongly remind one of the celebrated Dodo;" a remark to which he was guided by a sentence which he quotes from my Report on Ornithology (British Association Reports, 1844, p. 189), stating that Mr. Titian Peale "is said to have discovered a new bird allied to the Dodo, which he proposes to name *Didunculus.*" But Mr. Gould never stated that the *Gnathodon* (or *Didunculus*) was actually allied to the Dodo, and no one in this country had ventured to assert the affinity of the latter bird to the Pigeons, until, in the end of 1846 or beginning of 1847, I succeeded in convincing several naturalists that this affinity was real. Mr. Gould has politely informed me that a short time previously to the meeting of the Association "Dr. Melville showed me the dissected head of the Dodo from Oxford, together with skulls of several species of *Columbidæ,* when their similarity of form was so apparent that I became a convert to its Columbidine affinity."

[1] From the Hist. de l'Acad. Roy. des Sc. 1776, p. 37, it appears that Pingré published, or at least wrote, a relation of his voyage, in which he speaks of Solitaires, but I can find no notice of any such work among the published biographies of Pingré.

[2] The latitude of Pingré's observatory was 19° 40′ 40″ S., its longitude 4^h 3′ 26″ (or 60° 51′ 30″ E.) of Paris.

[3] "Birds of Australia, part xxii. Description of the *Gnathodon strigirostris*: the bird which its discoverer, Mr. Titian Peale, supposed to be allied to the Dodo, and proposed to name *Didunculus,* which was first described by Sir W. Jardine under the name of *Gnathodon strigirostris,* and which Mr. Gould regards as being most nearly allied to the family of *Columbidæ.*"

PART II.

OSTEOLOGY

OF THE

DODO AND SOLITAIRE.

BY

A. G. MELVILLE, M.D. EDIN., M.R.C.S.

INTRODUCTION.

In our efforts to determine the affinities of an extinct or fossil bird, by comparison of its osseous remains with the same parts in existing forms, we must be on our guard against relying too implicitly on the affinities which appear to be indicated by an incidental similarity in absolute size of the things compared, overlooking the more important elements for guiding us to a correct conclusion, namely, correspondence of general form and minute configuration.

Having obtained an approximate idea of the affinities by a comparison rightly instituted, we should next enquire whether the existing species of the type to which it has been referred afford a range in the form and relative proportions of important homologous parts, sufficiently wide to allow of its anomalies being admitted within the limits of the probable variations of the type.

The too frequent disposition to discern in each newly-acquired form, recent or extinct, one of those links between now dissevered groups of animated beings, which, from the imperfect nature of our conceptions we suppose to have been created, may lead the most truthful observer into error in determining its proper rank. The progress of discovery has indeed added members to some apparently defective families, but all attempts to fuse great conterminous groups together, have only more clearly illustrated the fundamental unity of organization, without destroying the multiplicity in that unity.

As in Mammals, the cranium with its dental armature is the part of the skeleton from which the Palæontologist derives the most certain indications as to the position of an extinct species; so in Birds, the same segment of the osseous frame-work is that which preserves the typical characters, notwithstanding such alterations in other parts as may even annihilate the power of flight, that almost universal characteristic of the class. The variations in the number, size, and pattern of the teeth in Mammals, denoting essential differences in the nature of the food selected, are parallelled in birds by modifications in the form, size and relative proportions of the beak, and its horny sheath.

The force and extent of the movements of the mandibles have an essential relation to the nature of the food, and the resistance to be overcome in its prehension. Hence the depth of the muscular fossæ, and the height of the ridges giving attachment to the muscles of mastication, cannot but convey to us valuable information, which should further be correlated with that resulting from the indications of the amount of movement of the head on the trunk. The form of the palatine bone especially deserves attention, from its giving attachment to one of the principal muscles employed in mastication, and moreover bounding the posterior nares and subocular cell. Unfortunately this bone is generally deficient in fossil crania.

The shape of the tympanic bone, and more particularly that of its inferior articular surface, are useful guides to classification. Much value is also to be attached to the form and position of the prefrontal (*lacrymal*, of authors), and to the circumstance whether it be anchylosed to the cranium, or separate from it; to the form and size of the posterior nasal fissures; to the presence or absence of the vomer, and of the ossified septum narium.

The general pneumaticity of the cranium, and the ratio in which the several elements participate in that property, furnish less distinctive characters; the development of pneumaticity depending on many variable conditions.

In the former part of this work the views expressed on the affinities of the Dodo by various distinguished zoologists and anatomists, are given at length; of these, Professor Owen alone had the opportunity of studying the evidence furnished by the foot, which led him to regard the Dodo as an extremely modified form of the Raptorial order. In the catalogue of the fossil remains of Mammalia and Aves in the collection of the Royal College of Surgeons, published in 1845, apparently before he had seen the dissected foot, the Dodo is placed among the Cursorial, or Struthious birds, from some vague resemblances in the cere and advanced nostrils, to the corresponding parts in different members of that limited group.

The merit due to Professor Reinhardt, who from the evidence afforded by the mutilated cranium in the Göttorf Museum, assigned to the Dodo, thus bandied about, a final resting place among the Pigeons, has been freely conceded by his fellow-labourer, Mr. Strickland; who, however, from a minute and accurate comparison of the bones of the leg with those of other types, had arrived at the same goal, by a different, but equally certain path. The idea once attained served to elucidate the true relations of the cere and advanced tubular nostril, which had hitherto been misunderstood; the disappearance of the mandibular horny sheath was also readily explained by the facility with which it desquamates in other members of this group. Some learned ornithologists admit, that the correct interpretation of these external characters alone, might have led to the proper allocation of this strange and almost fabulous creature.

From anxiety to obtain the fullest information, application was made to Mr. Duncan, Keeper of the Ashmolean Museum, for permission, which was liberally granted, to remove the integuments from the left side of the head of the Dodo, where they were most decayed, and the requisite dissection was judiciously performed by the Reader of Anatomy, Dr. Acland. During this procedure, the leading points of resemblance between the cranium and that of the Pigeons were pointed out by Mr. Strickland, who has kindly associated the writer with him, in the task of describing the remains of this extinct form and its affine, the Solitaire.

My testimony, hence, is that only of an impartial observer with no hypothesis to defend, and who claims no share in the merit due to those who have succeeded in restoring the Dodo to its proper rank.

THE

NATURAL HISTORY

OF THE

DODO, SOLITAIRE, &c.

PART II.

CHAPTER I.

Osteology of the Dodo.

(Plates VIII., IX., X., XI., and XII.)

The skull of the Dodo is larger than that of any existing raptorial bird, and greater though shorter than that of the Albatross; its ratio to that of the Goura and Treron will be seen by a glance at Plate X.

The skull is remarkable not only for its great absolute and relative size, but also, for the abbreviation of the cranium, whose length is to that of the upper mandible as one to two, and for the sudden rise of the frontal region above the compressed upper mandible; the skull hence assumes, as it were, the form of a mallet, the head of which corresponds to the cranium, while the core, or bony termination of the mandibles, acts as a counterpoise.

The shortening of the cranium is due to the small relative size of the eyes, and the consequent retrogression of the ethmoidal fossæ, and atrophy of the proper interorbital septum.

The elevation of the frontal region above the level of the upper mandible, is produced by a sudden expansion of the pneumatic diploë, tilting up the extremity of the mesial process of the premaxillary, and the body of the nasal on each side, at an angle of 45°; while the abbreviated frontal is raised into a broadly rounded interorbital eminence.

There is a similar development of the diploë, though in a less degree, in the Goura. The rise of the frontal region is in some Pigeons more abrupt than in the Dodo, but is owing to a different cause; namely, the great size of the orbit, and the relative slenderness of the bill.

The upper mandible, viewed from above, presents on each side a shallow excavation extending from the core to the base of the maxilla; the upper edge of the ramus of the lower jaw forms the chord of the concavity, which lodges the curved tubular nostril. This characteristic appearance is owing to the great compression of the lateral beams of the mandibular apparatus towards each other, by which they are, as it were, forced almost into contact beneath the upper stem; their height being thus increased at the expense of their breadth; while their oblique bases diverge towards their upper or terminal angles, and each beam resumes, so to speak, its original thickness.

The length of the skull, measured from the upper border of the foramen magnum to the apex of the mandible, is 8 inches $2\frac{1}{2}$ lines; its breadth, a little in front of the post-orbital process, is 3 inches $8\frac{1}{2}$ lines; the greatest elevation of the cranium is 2 inches 5 lines. The extreme length of the lower jaw is 7 inches 9 lines, and its span 2 inches 10 lines.

On a more minute examination, the skull of the Dodo will be found to present the typical characters of that segment in the Columbidæ, which are:—

I. A feebly uncinated *upper core*; a character which at once distinguishes the Dodo from the Vulturidæ on the one hand, and Cathartes on the other.

II. An external *nasal fissure* extending from the base of the core, as far as, or beyond the resilient hinge formed by the upper beam of the mandibular apparatus at its junction with the cranium; in all raptorial birds, the major part of the body of the nasal is placed in front of that line; while in Pigeons the body is abbreviated and rises high on the frontal slope, the divergence of its limbs exposing to view, in certain genera, the turbinated ala of the ethmoid. The rasorial genus Pterocles presents a similar character; hence it is not distinctive of the Columbidæ.

In the Vulturidæ, the nasal scale is ossified to support the horny cere, and the nostril opens anteriorly by a narrow vertical orifice; while in the Dodo, the elongated lanceolate nasal fissure extends to the foot of the frontal protuberance.

III. The elevation of the base of the *maxillary bone* to meet the expanded foot of the abbreviated ecto-nasal limb, and the obliquity of the zygoma, which must descend as it retrogrades from the junction of these bones, to the level of the inferior articular surface of the os quadratum. The maxillary in Pigeons is subpyramidal with a triangular section; the apex extending forwards, like a splint, on the inner side of the lateral process of the premaxillary; the external surface slopes obliquely upwards and outwards from the palatine aspect, and is more or less tumid; the angle which it forms with the inner concave facet is united to the pyramidal foot of the ecto-nasal limb behind, the termination of the lateral premaxillary process being wedged between them anteriorly. The ecto-nasal limb passes upwards and backwards from the upper angle of the base of the maxilla; the inner edge is prolonged into the antral plate, and is separated by a groove, on the floor of which occurs the pneumatic foramen, from the terminal border of the external surface, which ascends obliquely backwards, its upper angle passing into the slender zygoma.

The mandible thus presents a subtriangular surface, of greater or less extent, and more or less tumid, which is covered by the palatal membrane; it is continued backwards by the external fibrous wall of the subocular cell, extending from the root of the zygoma to the free outer margin of the palatine crest. The surfaces of opposite sides, separated mesially by the posterior nares, form a wedge subsiding anteriorly towards the nasal orifice, and descending between the rami of the lower jaw, which are so curved that their convexity mounts into the obtuse angle formed between the zygoma, and the lower margin of the lateral facet of the maxilla; which is indicated in the Dodo by the upper caruncular ridge, separating the palatal mucous membrane from the cere, and extending forwards to the lower angle of the nostril. In those grallatorial and aquatic birds, as the Ibis, Spoonbill, and Albatross, which have a similar arrangement, the margin of the upper mandible overhangs that of the lower; and in the Albatross the posterior part of the dentary bone is lodged in a deep groove, between the palatine wedge and the acute margin of the mandible.

IV. The absence of an ossified *vomer* separating the posterior nares; this is also generally deficient in the Gallinæ; in the Vulturidæ it exists in the form of a narrow lanceolate plate, but is wanting in Cathartes.

V. The *septum narium* is generally membranous in Pigeons; it exhibits however traces of ossification at its attachment above, in the *Calœnas nicobarica*, and *Lopholœmus antarcticus*; in the Vulturidæ it is wholly ossified, a small perforation only existing in certain species; in Cathartes it exists, although reduced in length, by the removal of its anterior part in the formation of the common nasal vestibule. In the Dodo it is completely membranous.

VI. The form of the *palatine bone* in Pigeons is characteristic, and differs from that in the Rasores, in the presence of the horizontal plate or crest, which affords an increased surface for the origin of the internal pterygoid muscle; and of the descending palatal process, which supports the fold of mucous membrane forming the lateral boundary of the posterior nares. In the Vulturidæ, the crest is much broader, indicating the greater strength of the muscle arising from it; the sphenoidal plate is narrower from the unexpanded condition of the rostrum; the nasal process is much contracted longitudinally, whereas in Pigeons it extends forwards along the inner margin of the palatine stem, to near its attachment; the palatine process is less elongated, and the inflected portion of it, in Pigeons, is entirely absent; the palatine stem is straight in Vultures, arched with the concavity inwards in Pigeons. In Cathartes the stem is also curved; the nasal process is more extended than in Vultures, but less elevated than in Pigeons; the crest however indicates the raptorial character by its great breadth.

We shall afterwards see how the form of this bone, in the Dodo, is modified by the compression of the mandible and the abbreviation of the rostrum, without departing from the Columbine type.

VII. The shape of the inferior articular surface of the tympanic bone, although it varies in different genera of the Columbidæ, is distinguished from that in the Vulturidæ, by the greater transverse diameter of the internal, and by the greater breadth of the external condyle which is flat, or slightly convex, and subcircular. In the Vulturidæ and Cathartes, the latter is narrow and sigmoidal; convex in front, concave behind longitudinally, and rounded transversely. In Cathartes the inner condyle is grooved at the base externally; and the trochlear ridge is more oblique than in ordinary Pigeons. In the Dodo, its form is similar to that in the typical genera of the Columbidæ, and differs from that in the large Vultures, with which, from a correspondence in absolute size, it may be more readily compared.

The absence of the posterior superior condyle in the typical Rasores, and its presence in Pterocles, approximates, so far, this aberrant genus to the Columbidæ.

VIII. The subtriangular body of the *prefrontal* is dove-tailed between the nasal bone and antorbital process of the frontal, which advances along its outer edge to the lacrymal groove; in the adult it is anchylosed to these bones above, and internally to the highly developed alæ of the ethmoid; the prefronto-ethmoidal fissure being in most Pigeons wholly obliterated. In Goura, a slender style separates its inner margin from the nasal, so that it is inserted by gomphosis, into a deep semi-elliptical notch on the broad antorbital process. It is not subject to removal by maceration, or such forces as would almost inevitably break off the upper mandible; and its occurrence in the fractured cranium of the Solitaire, may be regarded as presumptive evidence of the Columbine affinities of that extinct form.

In Pterocles, the prefrontal is anchylosed, but I have not been able to ascertain its relation to the antorbital process; from the narrowness of its frontal aspect, it is not probable that this process extends along its outer margin. In the typical Rasores, the prefrontal is free, and projects greatly outwards; its inferior process is reduced to a slender curved style; and the alæ of the ethmoid are wanting, while in Pterocles they are greatly developed, and the prefronto-ethmoidal fissure is obliterated. The prefrontal is unanchylosed, even in the adult, in all raptorial birds, except the aberrant genus Cathartes; the free external angle supports the os superciliare; the prefronto-ethmoidal fissure is large and persistent; and the antorbital process forms only a slight angular separation, between the shallow notch lodging the apex of the prefrontal, and the deeply concave superciliary margin, which sweeps rapidly outwards and downwards to the post-orbital process.

In Cathartes, the prefrontal is firmly united to the cranium; the supra-orbital membrane is completely ossified, and gives increased breadth to the forehead. The olfactory foramen opens into the apex of the infundibular turbinated ala of the ethmoid; the inferior ala is anchylosed to the prefrontal below, but the prefronto-ethmoidal fissure remains.

IX. The size of the *crotophyte impression*, although variable in different species, according to the resistance to be overcome, is very minute when compared with that in the Vulturidæ, or even Cathartes; in the Dodo it is exceedingly small, and is not compensated by an increase in the area of the internal temporal surface.

X. The great extent of the *digastric impression* in Pigeons and in the Dodo, is well contrasted with its small size in raptorial birds. The Rasores in this respect, as might be anticipated, resemble the Columbidæ.

XI. The presence of a single *mesial supra-occipital aperture* above the foramen magnum, for the transmission of a vein, which arises from the muscles of the neck, and joins the posterior cerebellar sinus. Among the Raptores, it occurs in some Owls, but I have not seen it in any other family of birds. Its co-existence in the Dodo with other indications of affinity to the Columbidæ, shows the value of apparently trivial characters in determining the position of an anomalous form.

XII. The general *pneumaticity* of the cranial vault is greater in Pigeons, and the prefrontals and sphenoidal rostrum are usually much more expanded than in the Vulturidæ and Cathartes. In these respects the Dodo resembles the Columbidæ, and differs remarkably in the bullose appearance of the prefrontal, and in the breadth of the rostrum, from the typical raptorial birds. The Pterocles also approaches the Columbidæ in these characters.

XIII. In the *lower jaw*, the curvature of the rami; their union at a more or less angular, short and ascending, symphysis; the separation of the dentary, and, in some cases to a late period, of the opercular elements; the presence of the interangular foramen in certain genera; the large triangular digastric, or basal, facet; the small area of the temporal and pterygoid impressions; and the differences in the form of the articular surface, corresponding to those already alluded to, in the inferior surface of the tympanic, distinguish the lower mandible, in the Columbidæ from that in the Vulturidæ and Cathartes: in the latter, however, the lower jaw is more curved than in the less aberrant Raptores. We shall afterwards see how these important differences are repeated in the Dodo. The development of the basal angles of the digastric facet into the posterior and internal angular processes, so characteristic of the typical Rasores, is observed in Pterocles.

The family characters of the skull in the Columbidæ, just enumerated, are derived from the consideration of parts, important either in a physiological, or morphological, point of view. One or more of them may be absent in aberrant members, or be common to different types; but the whole, or a majority of them, occurring in the skull of an extinct form, would justify us in assigning to it a place among this interesting and extensive group.

Before proceeding to a more minute description of the skull of the Dodo, and to a comparison of it with that of other Pigeons, we may recapitulate shortly, those important differences which warrant us in restricting such comparisons to the members of the Columbine group.

The skull of the Dodo differs from that of the Vulturidæ, in the relatively small and feebly uncinated core; in the elongation of the external nasal orifice, and absence of the ossified scale; in the great relative size of the maxillary bone; in the obliquity of the zygoma; in the form of the mandibular surface of the tympanic; in the form of the palatine bone; in the absence of the ossified septum narium; in the absence of the vomer; in the form, and minute configuration, of the lower jaw; in the anchylosis of the prefrontal, and obliteration of the prefronto-ethmoidal fissure; in the greater breadth of the interorbital region, and absence of the os superciliare; in the small area of the crotophyte impressions, and the great relative size of the digastric surface; in the existence of the mesial supra-occipital foramen; in the great pneumaticity of the cranium; in the ratio of the upper mandible to the cranium; in the retrogression of the ethmoidal fossæ; in the absence of the frontal protuberance; and of the lateral excavations of the upper mandible. Such, then, are some of the more important and characteristic distinctions; without entering into superfluous details, regarding the differences in the minute configuration of similar parts, as these may vary even in the same group.

The Cathartes resembles the Dodo, in the absence of the ossified vomer; in the anchylosis of the prefrontal; in the retrogression of the ethmoidal fossæ; in the breadth of the interorbital region; and in the curvature of the lower jaw. It differs, however, in other more important characters, common to it with the typical Vultures, and is peculiar in possessing the nasal vestibule, characteristic of the Cathartine modification of the raptorial type.

From the typical Rasores, the Dodo differs, in the elongation of the external nasal orifice; in the greater development of the maxilla; in the obliquity of the zygoma; in the greater complexity of the palatine bone; in the double mastoid condyle of the tympanic; in the absence of the posterior and internal angular processes of the basal facet of the lower jaw; in the anchylosis of the prefrontal, and great development of its inferior process; in the presence of the alæ of the ethmoid; in the retrogression of the ethmoidal fossæ; in the great pneumaticity of the prefrontal, and of the sphenoidal rostrum; and in the absence of the mesial supra-occipital foramen.

The skull of Pterocles resembles that of the Dodo in the same degree as it approaches the type of the Columbidæ.

From the Insessores, the Dodo is at once distinguished by the form of the palatine bone; by the absence of the vomer; by the elongation of the external nasal fissure; by the obliquity of the zygoma; and by the relation of the antorbital process to the prefrontal.

It would be useless to state the essential differences between the skull of the Dodo, and that in the different families of the Grallatorial and Natatorial orders, as no one is likely to suppose that it has any affinity with either of these groups.

I now proceed to describe the skull of the Dodo in greater detail.

The *posterior* subelliptical facet of the cranium, is formed by the occipital bone; its greater diameter is transverse, and measures two inches and eight lines and a half; and its lesser, one inch and seven lines and a half. It presents an upper crescentic segment, with a vertical plane, embracing in its concavity a lozenge-shaped surface, which inclines obliquely downwards and forwards at an angle of 125°

The upper convex margin corresponds to the supra-occipital ridge, continued on each side into the convex incurving border of the paroccipital process, which projects outwards, forming the posterior wall of the tympanic cavity. The infra-occipital ridge, forming the central moiety of the concave edge, overhangs the recess perforated by the foramen magnum, like the dripstone of a Gothic arch; a line drawn from its corbal-like origin outwards, on each side, to the inferior angle of the paroccipital process, indicates the remainder of this boundary; along which the vertical supra-occipital surface is broadly rounded off into the rhomboidal fossa, occupying the lateral angle of the lozenge. This depression is bounded externally by the lower sharp edge of the internal wall of the tympanic cavity, arching from the paroccipital angle to the posterior border of the basilar pyramidal protuberance, which projects vertically downwards; a rough prominent ridge, notched in the centre, ascends, inclining inwards and backwards along its internal edge, and with its fellow forms the anterior boundary of the occipital facet, indicating nearly the division between the sphenoid and occipital bones. The supra-occipital plate presents, in the centre, a triangular, broad and depressed cerebellar elevation; the truncated apex is on a level with the infra-occipital ridge, a line above which it is perforated mesially by a short canal, half a line in diameter, opening internally immediately within the upper margin of the foramen magnum: a slight crest traverses the median line, becoming more apparent as the convexity subsides towards the base, the angles of which extend to the most concave part of the supra-occipital ridge; below it enters the furrow leading to the orifice just mentioned. The large oblong surface, external to the central protuberance on each side, is divided into two subequal portions, by a convexity directed downwards and inwards from the origin of the superior occipital ridge to that of the inferior, it corresponds to the semicircular canals within; on the left side it is widest and most prominent superiorly, and subsides towards the lateral venous groove; the floor of the inner segment is slightly elevated towards the supra-occipital ridge, but on the right side it is raised into a triangular convexity, more prominent than the canalicular elevation, to which it is parallel, separated only by a slight digital impression. The lateral venous groove passes obliquely inwards and upwards, above the infra-occipital ridge, and terminates at the margin of the cerebellar protuberance, in a short oval canal perforating the cranium, two lines and a half above the foramen magnum; and opening upwards, internally, at the margin of the cerebellar fossa; it transmits the lateral venous sinus. The groove contracts in the centre of its course, and is there covered by a narrow osseous bridge on the left side; a small canal lodging a muscular vein opens downwards into the internal segment. Externally, it curves round the origin of the infra-occipital ridge; a narrow tract, one line in breadth, separating its termination from the groove for the bulb of the jugular vein, which the lateral sinus joins.

The area of the strongly pitted muscular impression, on each side, is elliptical; its inner angle is prolonged inwards across the base of the cerebellar protuberance; the outer occupies the upper part of the paroccipital process, the lower portion of which is smooth. In its centre it extends from the venous groove to the supra-occipital ridge, and the surface is increased by the elevations already described. The cerebellar eminence is smooth and polished. From the notch between the paroccipital process and mastoid, inwards for half an inch, the digastric and occipital impressions are separated only by a smooth convex edge; more internally, a prolongation of the parietal tract intervenes, the apex extending to the canalicular elevation; where the supra-occipital ridge originates.

This ridge is broad and rough externally; it ascends on each side, becoming narrow and rounded, following the undulation of the surface to the angle of the cerebellar elevation; from thence it descends to near the mesial line, over which it arches: it presents a slight notch at the upper and inner angle of the canalicular convexity, from which a groove leads outwards and downwards to a canal perforating that eminence, and traversing the cranial diploë to open on the lateral facet below the superior pneumatic foramen; it transmits a vein from the integuments of the cranium. The supra-occipital ridge is defined by the subsidence of the posterior surface, not by its elevation above the parietal tract.

The convex margin of the paroccipital process increases in breadth inferiorly, and is rough and flattened, giving origin to the Biventer maxilla muscle; a strong ligament passing from its lower angle, forwards and inwards to the apex of the basilar facet of the ramus of the lower jaw, is still present.

The infra-occipital ridge is broad, rough, and prominent externally, the lateral venous groove bending round its origin; the roughness subsides internally, as it passes into the cerebellar eminence. The foramen magnum has an ovate form, subangular above; it is six lines and a half high, and nearly five lines broad, inferiorly; the vestibular elevation in the interior of the cranium, projects on each side into the area of the foramen beneath the centre, rendering it fiddle-shaped, but to a greater extent on the left than on the right side; it is four lines in width where thus constricted. Its margin is separated from the infra-occipital ridge all round by a groove, which widens below from the inclination forwards of the plane of the foramen; its breadth is one line and one-third above mesially, and three lines and a half below. This recess lodges the great posterior cerebellar sinus, which discharges itself by a large branch perforating the posterior occipito-atlantal ligament, to form the bulb of the internal jugular vein.

The occipital condyle is subpedunculated; its axis is directed downwards and backwards, so that its posterior surface, above, is nearly in the same vertical plane as the margin of the foramen magnum; the articular surface is separated by a groove from the peduncle laterally, but is continued on it inferiorly, indicating a considerable amount of downward flexion of the head on the neck; its form is subhemispherical, flattened, and notched above by the prolongation of the fossa for the medulla oblongata; its posterior surface is marked by a faint median vertical groove; its height is three lines, and its breadth four lines.

The lateral basilar fossa presents posteriorly the shallow oval concavity for the bulb of the jugular vein; it is most distinct on the right side; its inner angle is separated from the groove surrounding the foramen magnum by a convex ridge, it is directed outwards and forwards to the foramen lacerum posterius, grooving its inner edge; and is four lines and a half long, and two and a half broad. The outer boundary of the deep elliptic fossa, four lines and a half long and two broad, common to the posterior lacerated and carotic foramina, is formed by the inner wall of the tympanic cavity, the lower sharp edge of which, concave inferiorly, extends as already indicated from the paroccipital angle to the pyramidal protuberance; its inner margin is more deeply concave than the external, thin and notched behind, thick and rounded in front. The fossa is divided by a roughened transverse convexity, forming the floor of the vestibule leading to the foramen ovale and f. rotundum; the foramen caroticum, transmitting the internal carotid and its accompanying sinus, leads forwards and inwards from the anterior angle; while, from the posterior, passes upwards, curving forwards round the vestibule just mentioned, a canal, which transmits an artery and accompanying vein, with the glosso-pharyngeal and sympathetic nerves; its inferior orifice corresponds in part to the foramen lacerum posterius of mammals; its outer wall is perforated, a line above its margin, by a rounded aperture leading to a broad groove on the base of the paroccipital process anteriorly; it transmits the venous sinus of the membrana tympani to the internal jugular vein. The condyloid foramen for the passage of the hypoglossal nerve, perforates the base of the peduncle of the occipital condyle, one line and one-third external to the foramen magnum; one line and a half from it, towards the carotic canal, is the large aperture, transmitting the pneumogastric and spinal accessory nerves. Two minute apertures separated by an osseous line, and probably giving exit to small venules, perforate the lower part of the interspace between the condyloid and pneumogastric foramina on the left side; on the right side they are a line and a half apart.

The lateral fossa is bounded in front, by the posterior convex surface of the triangular basilar pyramid, whose elongated inner edge extends inwards, sloping backwards to the groove, uniting the inner and anterior angles of the fossa of each side beneath the occipital condyle. The apex and narrow outer surface are rough: a scabrous tubercle is developed at the foot of the inner edge, separated from its fellow by a smooth mesial notch. The breadth of the supra-occipital plate, between the mastoid notches, is two inches eight lines and a half; from the median line obliquely to the mastoid notch, it measures one inch eight lines and a half; its height, mesially, is nine lines and a half; and its thickness towards the occipital ridge is five lines and a half above, and one line and a half below. The distance between the inferior angles of the paroccipital processes, is one inch six lines and a half, and its depth from the upper margin of the foramen magnum, ten lines and a half; from the mesial line obliquely to the angle of the paroccipital process, is one inch two lines. The longitudinal diameter of the inferior segment is eleven lines and two-thirds; and the span of its lateral boundary is eight lines and a half. The span of the anterior margin between the apices of the basilar protuberances, is eight lines and a half. The basi-occipital is three lines and a half thick.

The *inferior* facet of the cranium exhibits, on each side, the great orbito-temporal excavation, separated by the short but broad and tumid sphenoid, which expands to join the basi-occipital posteriorly; it is most constricted at the junction of its posterior and two anterior thirds, opposite the foramen opticum; it increases in breadth, as it advances, to where it is joined by the inferior ethmoidal ala on each side; it then rapidly becomes narrower, and forms a compressed plate projecting forwards into the inferior nasal fissure.

The inflated lower ala of the ethmoid coalesces externally with the bullose inferior extremity of the prefrontal; which, from the compression of the cranium anteriorly, is prolonged forwards; a deep narrow notch, leading into the olfactory fossa, separating it from the rostrum. The greatest width of the sphenoid, between the outer margins of the tympanic tubes, is one inch, eleven lines and a half; its extreme length is two inches, six lines and a half; where most constricted it is six lines and two thirds broad. The distance between the outer surfaces of the inferior extremities of the prefrontals amounts to one inch, eight lines and a half; the thickness of the sphenoid at the foramen opticum is four lines and a half.

The lozenge-shaped concavity on the sphenoid, for the insertion of the fleshy fasciculi of the *Recti capitis antici* muscles, is deepened posteriorly and laterally by the anterior surface of the basilar protuberance, which is directed, on each side, inwards and backwards, sloping outwards as it descends to the apex; anteriorly and laterally a slight ridge separates it from a venous impression on the side of the sphenoid. The anterior angle projects in the form of a small, free, triangular plate, curving downwards beneath the transversely ovate common orifice of the Eustachian tubes; the posterior corresponds to the median notch between the inner angles of the basilar protuberances. This surface is distinctly pitted on each side of an irregular raphe: and its antero-posterior diameter is eight lines.

A groove, two lines broad, running forwards from the orifice of the Eustachian tubes, impresses the sphenoid where most constricted, and ceases after a course of four lines with a rounded termination; its edges are sharp, and separate it from the lateral venous impressions; it is lined by the sinus leading to the Eustachian tubes.

Anteriorly the sphenoid is raised into an obtuse median ridge, between the flattened oblong surfaces on which the palatine and pterygoid bones glide, in the movements of the upper mandible.

The inferior aspect is concave upwards, deepest opposite the foramen opticum, declining in front to the lowest part of the rostrum, and behind to the under surface of the occipital condyle, which is in the same horizontal plane as the former.

The *lateral* aspect of the cranium is occupied by the large orbito-temporal fossa, which presents the form of an irregular, four-sided pyramidal excavation; the floor and inferior wall of which are removed. Its different surfaces converge to the optic foramen; the posterior is of less height than the anterior, the upper descending as it retrogrades.

The anterior subconcave wall, is formed in its anterior moiety by the prefrontal, and by the united turbinated and inferior alæ of the ethmoid; posteriorly it is constituted by the enormously thick, but contracted, interorbital septum; it slopes gradually inwards and backwards, and above is rounded off into the roof of the orbit. A line drawn from the post-orbital process to the optic foramen, divides the orbital from the narrow, depressed temporal fossa, which inclines forwards, as it descends inwards. The posterior triangular surface is broken by the projection of the anterior wall of the tympanic cavity, it slopes backwards and outwards; its upper margin extends to the post-temporal process of the mastoid; the lower is horizontal and directed to the inferior angle of the paroccipital process.

The roof of the orbit is formed by the frontal, and by the ali-sphenoid, for a short space anterior to the temporal fossa; a line drawn from the optic foramen, through the foramen ovale, to the posterior tympanic articular facet, will indicate the lower margin of the ali-sphenoid. The inferior boundary of the mastoid corresponds to a line drawn from the notch between it and the paroccipital process, through the superior tympanic aperture to the inner angle of the root of the post-temporal process, where it comes into apposition with the external border of the ali-sphenoid; thence the suture passes forwards, inclining upwards to the post-orbital process. The division between the sphenoid and the ex-occipital follows the course of the canal which runs behind the fenestra ovalis; its upper angle is anterior to the inferior tympanic articular surface, which is developed on the ex-occipital; below it passes internal to the Eustachian tube, cutting through the elliptic fossa, common to the foramen caroticum and f. lacerum posterius, and lastly bends transversely inwards, intersecting the foot of the basilar protuberance. The diminished area of the interorbital septum, which is only about six lines in diameter, is remarkable, and is due to the small size of the eyes, which are amply protected by the great outward projection of the roof of the orbit posteriorly. The proper septum is reduced to the small space, intervening between the base of the olfactory fossæ and the interval separating the foramina optica, in the antero-posterior diameter; it is encroached on above by the expanded frontals, and below by the inflated rostrum. From the abbreviation of the cranium, and consequent shortening of the frontal, the orbital vault is relatively very small; it is bent down abruptly anteriorly, nearly at right angles, and, as it were, pressed backwards; the angle of flexure corresponding to the supra-orbital notch, from which the roof increases in breadth as it retrogrades obliquely downwards.

A line drawn from the supra-orbital to the temporal notch would cut off an elongated triangular segment; the hypothenuse corresponding to the convex, thick, and rough supra-orbital margin, and the base to the post-orbital process. The great breadth of the interorbital region, which is continued backwards diminishing very gradually to the mastoid notch, and the flattening down, as it were, of the roof of the orbit behind the eye; together with the great elevation of the forehead above the surface of the mandible, and its contraction in front of the supra-orbital notch, are remarkable peculiarities in the head of this extinct form. The roof of the orbit is arched transversely, but more flatly concave longitudinally than the anterior portion of the orbital vault; the greatly increased expansion of the diploë of the frontal internally, causes its surface to descend rapidly into that of the interorbital septum; while from the retrogression of the olfactory fossæ, the anterior wall slopes very gradually backwards. A line drawn from the inferior extremity of the prefrontal to the post-orbital process, measuring one inch and seven lines, ascends obliquely backwards at an angle of 45°; and a plane extended inwards from it to the optic foramen would limit the orbit posteriorly and inferiorly. The depth of the vault from the supra-orbital notch, is one inch one line.

The *foramen opticum*, situated at the apex of the triangularly pyramidal orbital fossa, is equidistant from the anterior and posterior surfaces of the cranium, and from the supra-orbital and mastoid notches; its circular contour is notched above by a vascular groove, and its relatively small diameter is two lines and a half. Its floor is four lines and a half from the basilar surface, and its roof one inch ten lines and a half beneath the highest point of the frontal protuberance. The anterior edges of the foramina of opposite sides, are separated by an interval of six lines and two-thirds, corresponding to the broad posterior border of the interorbital septum, which is convex transversely, and concave vertically.

The *ant-orbital foramen*, for the transmission of the ophthalmic branch of the trigeminal nerve and accompanying vessels, is eight lines distant from the foramen opticum, in a direction forwards and upwards, and midway between it and the ant-orbital process. Its form is irregularly transversely ovate; a ventricose projection on the left side, encroaches on its area inferiorly. The posterior expanded border of the inferior ethmoidal ala, is three lines and a half broad. The prefronto-ethmoidal fissure is obliterated; a narrow, tripartite chink, slightly wider in the centre, alone remaining; the inner branch of which probably corresponds to the notch between the turbinated and inferior alæ, the evasation of the latter to form the large olfactory fossa being the cause of the disappearance of this fissure, which in other birds transmits the superior diverticulum of the subocular sinus; the upper indicates the union of the prefrontal to the turbinated ala; from its extremity an interrupted groove, more distinct on the right side, is directed downwards and outwards at an acute angle, to the upper margin of the lacrymal groove, defining the antorbital process as it runs along the outer margin of the body of the prefrontal.

The tumid prefrontal coalesces below with the inferior ala, and is three lines and two-thirds wide; its broad outer convex surface, beneath the antorbital process, presents the lacrymal groove, more depressed at its lower margin; it runs forward, inclining inwards, to the outer margin of the olfactory fossa, and is eight lines in length, and three and a half broad; in the inner part of its course, it rests on a quadrate process of the prefrontal, which projects beyond the level of the anterior margin, and comes into contact with the convexity of the turbinated ala; this projection is separated by a notch, from the prominent inner angle of the body of the prefrontal above, and below from its inferior extremity; which is slightly flattened and roughened opposite the zygoma, with which it would probably come into contact, in the great downward flexion of the upper mandible. The ant-orbital process is thick and rough externally, and contracts in its anterior moiety, into a narrow style, whose apex is at the upper border of the lacrymal groove. From the supra-orbital notch a deep capillary fissure, with small lateral offsets, passes backwards and inwards, on the roof of the orbit for six lines; it probably lodged a small branch derived from a cutaneous artery. The superciliary margin is perforated about six lines in front of the post-orbital process, by two small foramina on the left side, but is notched on the right for the transmission of the supra-orbital arteries and veins to the scalp; internally they correspond to a groove running half an inch in front of, and parallel to, the posterior border of the orbit; it winds round a tumid projection of the diploë of the ala-sphenoid a little above the foramen opticum, which it enters, grooving its roof, and disappearing as it curves backwards. A second groove, for the nasal vessels, runs backwards and upwards from the ant-orbital foramen, to join the supra-orbital furrow above the prominence just mentioned. Numerous small apertures are seen along the course of these channels; a faint vascular groove runs from the prefronto-ethmoidal fissure to the centre of the nasal one; between the latter and the foramen opticum, and bounded laterally by the peculiar pneumatic bullæ, is a quadrate space, variously marked by vascular impressions.

The temporal fossa descends obliquely forwards, sloping inwards, and terminates inferiorly at a deep digital cavity, impressing the ali-sphenoid behind the optic foramen. It opens superiorly by the small narrow oblong temporal notch, five lines in depth, and three and a half in breadth; bounded in front by the short, thick, post-orbital process, slightly recurved at the apex, and behind by the post-temporal plate of the mastoid. The small crotophyte impression occupies the temporal gorge, and extends outwards, as the latter is broadly rounded off into the upper facet, in the form of a crescent, whose limbs extend on the triangular surface of the post-orbital process, and on the quadrantal post-temporal plate, which is traversed by a slight chord-like ridge. The internal temporal impression has the figure of a right-angled triangle; below, a narrow smooth tract separates its base from the surface for the *M. Levator ossis quadrati*; the undulated hypothenuse ascends forwards to the root of the post-orbital process, its upper third being separated by a faint ridge from the external impression; its surface subsides, anteriorly, about a line beneath the smooth and polished elevated area of the orbital fossa, the posterior rounded edge of which forms its anterior margin, which is seven lines long; its base is four lines and one-third; for a line above it, the surface is, as it were, scooped out.

A smooth tract, corresponding to the post-orbital vascular flexus, and leading to the *foramen rotundo-ovale,* occupies the remainder of the floor of the temporal fossa: it terminates below, at the convex projection of the *tympanic tube,* which runs horizontally outwards and backwards; its anterior wall is formed by a curved plate, which, after covering the inferior efferent pneumatic cells and the Eustachian tube, is attached to the outer margin of the sphenoid lozenge and basilar protuberance; behind which, it presents a deep triangular incision, bounded below by the thin, inferiorly concave edge; formed by its junction, at an acute angle, with the inner wall. Its orifice is in a line with the post-temporal process; the edge is deeply concave, the lower angle being prolonged into a sharp, slightly incurved styloid process. The tympanic convexity subsides internally towards the digital fossa, which is the deepest part of the lateral facet, containing in front the optic foramen, and behind, the subdivisions of the *foramen lacerum anterius,* perforating the thinnest part of the cranial parietes. The foramen for the transmission of the oculomotor and abducens nerves, occurs immediately behind the lower moiety of the optic foramen; it is longitudinally oval, one line and a half long, and one line high; it is directed obliquely upwards, and divided internally into three apertures by delicate osseous threads: the posterior one is the orifice of a canal, four lines and a half long, which lodges the sixth nerve; the two anterior give passage to the divisions of the third pair, and run into a vertical groove separated by a ridge from that which is continued upwards from the posterior orifice, both terminate on a level with the upper margin of the foramen opticum; at the apex of the anterior is situated the minute orifice of a canal, two lines long, for the transmission of the patheticus nerve; it is capable of admitting a fine bristle; a groove passes forwards and upwards from it. At the lower angle, between the optic foramen and that for the third nerve, is the aperture of a very slender canal, opening internally into the *sella turcica,* and probably conducting outwards, a twig from the ento-carotid artery. Above the tympanic convexity, and three lines and a half from the posterior border of the foramen opticum, is the rounded, sharp-edged aperture, one line in diameter, for the transmission of the ophthalmic branch of the trigeminal nerve; a groove, concave downwards, leads forwards across the patheticus foramen, and apparently enters the optic outlet, behind the supra-orbital furrow. The vertically oval orifice, *foramen rotundo-ovale,* giving passage to the second and third divisions of the fifth pair, occurs midway between it and the margin of the osseous tympanic aperture; it is one line and a half high, and one wide; an osseous thread separates from its base anteriorly, a minute foramen.

The digital fossa is separated by a slight ridge from a concavity on the side of the most constricted part of the sphenoid, beneath the foramen opticum; this depression is produced by the great trunk of the internal maxillary vein, it slopes inwards below, and is separated from its fellow by the narrow, sharp-edged gutter on the inferior surface, leading to the common orifice of the Eustachian tubes; its superior border is convex upwards; in the posterior triangular tract between it, and the edge of the digital fossa, are two foramina; the anterior and upper is the smallest, and longitudinally oval; the posterior and larger has the same shape, but its greatest diameter is at right angles to that of the former. A narrow band, one line in breadth, separates them, and presents two capillary apertures above, but supports below the minute, thin and flexible upwardly-curved style, representing the articular peduncle for the pterygoid; which does not exist on the right side. These foramina transmit communicating branches from the internal maxillary artery to the ento-carotid, with their accompanying venous sinuses; the canals to which they lead pass backwards in the septum between the Eustachian tube and the efferent pneumatic cells, and open into the carotid canal as it curves inwards to unite with its fellow.

The triangular surface for the attachment of the *M. Levator ossis quadrati,* occupies the digital fossa, extending round the posterior border of the foramen opticum; it is prolonged outwards over the tympanic convexity, surrounded by a smooth depressed marginal tract of the latter, corresponding to a vascular circle. The irregular tympanic excavation is bounded above by the mastoid, and behind by the paroccipital process. The post-temporal process of the mastoid is formed by a thick quadrantal plate projecting outwards with an inclination forwards, behind the temporal fossa; its outer edge is rough, and gives attachment to the external mandibular ligament, and viewed laterally projects downwards and forwards like a styloid process in front of the articular cavity for the anterior superior condyle of the tympanic. This cavity impresses the base and posterior surface of the post-temporal plate, thinning it internally; its antero-posterior and transverse diameters, are three lines and a half. The mastoid process is a longitudinally extended, low, obtuse and thick pyramidal plate, projecting downwards so as to conceal the large quadrate superior pneumatic foramen, internal to it and between the tympanic articular cavities; its external smooth facet is separated by a slight ridge from the small posterior one, which is grooved at its base, and separated by a notch from the paroccipital process; its inner surface is reticulate, and forms with the external a sharp edge. The inferior articular facet for the reception of the postero-superior condyle of the tympanic is oblong, three lines and two thirds long and two lines and two thirds deep, and composed of a smaller inner and a larger posterior segment, at right angles to each other; beneath its anterior extremity is the external lacerated wall of a canal which ascends from the foramen lacerum posterius, curving forwards round the tympanic tube; it passes into a groove, arching backwards to the lower angle of the upper pneumatic orifice, and terminating in a narrow canal which traverses the diploë at the floor of that orifice, emerges at the canalicular convexity on the occipital aspect. The paroccipital process presents a shallow groove at its base anteriorly, which widens as it ascends; its floor is cellular above, behind the inferior tympanic facet; below it curves inwards, and passes into a rounded orifice which perforates the outer wall of the canal just mentioned; the groove lodges the sinus of the membrana tympani, which transmits its blood to the internal jugular vein; its outer edge is undefined, the inner is sharp and gives attachment to the membrana tympani.

Internal to it is a shorter but deeper concavity, also exposing pneumatic cells beneath the inferior tympanic facet. In the mouth of the tympanic tube is seen posteriorly the vertically oval orifice of a short canal, leading inwards and slightly forwards to the foramina ovale and rotundum of the vestibule, which are separated by an oblique grooved bar; in front of it, is a large pneumatic orifice transmitting air to the diploë surrounding the labyrinth, and the oval orifice of the depressed basilar efferent pneumatic canal, passing forwards and inwards separated by a thin septum from the wide Eustachian tube: the efferent apertures from which open into the cells of the basilar protuberance, over which it passes converging to its fellow; the common orifice has already been described. The anterior wall of the tympanic tube is perforated by an aperture leading into the pneumatic canal. From the supra-orbital to the mastoid notch is two inches four lines and a half; between the opposite surfaces of the prefrontal, and of the paroccipital process inferiorly, is an interval of two inches; the anterior margin of the orbit is one inch six lines and a half deep, and from the temporal notch to the lower angle of the paroccipital, one inch five lines and a half.

The broad *superior* facet of the cranium, on the removal of the beak so as to expose the upper surface of the turbinated alæ of the ethmoid, presents a subhexagonal figure; the anterior border, corresponding to a line drawn between the anterior angles of the prefrontals, being only one half of the width of the posterior; and the antero-lateral margins about twice as long as the postero-lateral: behind the line of the resilient hinge, formed by the beak and the cranium—the cranio-facial line, at which the forehead rises abruptly above the level of the upper mandible, it has the same form, but the anterior and posterior, as well as the lateral edges respectively, approach to equality in length.

Its greatest breadth, corresponding to a line drawn between the lateral rounded angles, about six lines in front of the post-orbital processes, is three inches nine lines; anteriorly, it contracts gradually to the supra-orbital notches, where it is three inches wide; it continues forwards for half an inch of the same diameter, and then rapidly diminishes in width to the anterior edge, which is one inch six lines.

A little behind its greatest transverse diameter, it presents the deep temporal emarginations, and gradually contracts to the notches separating the mastoid from the paroccipital processes, where it measures, transversely, two inches nine lines. The median longitudinal diameter, from the cranio-facial line to the occipital facet, is two inches nine lines; from the anterior angle of the prefrontal to the most remote part of the occipital aspect, is three inches two lines.

This facet is formed, behind and centrally, by the confluent, short but broad, parietals; posteriorly and laterally by the mastoid, presenting the muscular impressions, and extending forwards, so as to enter into the composition of the post-orbital process (which in the Emeu and Bustard is formed by a separate element); anteriorly it is constituted by the abbreviated and coalesced frontals, which are raised by the sudden and great expansion of the diploë into a broadly rounded, interorbital protuberance. The wide semi-lunar notch, formed by their combined anterior edges, receives the bodies or frontal plates of the nasals, which are abruptly bent upwards at an angle of 45° with the plane of the upper mandible, and ascend high on the frontal slope to coalesce with the frontals, the sutures being obliterated: the nasals appear to be relatively much abbreviated, and to be almost, if not wholly, separated mesially by the broad triangular extremity of the nasal process of the premaxillary, which is wedged between them, being bent upwards in the same peculiar manner. The vacuity left between the nasal bone and the ant-orbital process of the frontal, on each side, is filled by the triangular body of the prefrontal; which is anchylosed externally to the ant-orbital process, the latter advancing along its outer edge to the lacrymal groove, as already indicated; internally it is separated from the ecto-nasal limb by a fissure, but its apex is anchylosed to the frontal plate of the nasal. On removing the beak, the broad, flat arch is seen, formed by the prolongation of the interorbital septum and the turbinated lamina passing out from it, on each side, and curving downwards to meet the prefrontal.

In the immature condition, the peculiar frontal protuberance of the *Dodo* would not be developed, and the cranium would present a gentle slope, descending from the vertex (which is somewhat in front of the coronal fontanelle, and corresponds internally to the most elevated part of the cerebrum), to the upper surface of the mandible.

The profile would hence resemble that in the skull of the *Calœnas*, &c., but would be relatively much shorter, from the abbreviation of the frontal: the length of that bone, and more particularly of its orbital segment, depends on the extent traversed by the peduncle of the olfactory nerve, ere it terminates in the proper nerve-filaments distributed to the sense-capsule (*ethmoid*). It protects, as the upper segment of the fronto-neural arch, not as the lateral moiety of a divided spine, the anterior extremity of the cranio-vertebral tube, and is supported below by the interorbital septum, or centrum, of the frontal vertebra; which is excavated and reduced to a thin vertical plate, by the fossæ for the reception of the eyeballs and their appendages. In the *Dodo*, from the small relative size of the eyes, the interorbital septum assumes more of the ordinary characters of a centrum, and the olfactory capsules retrograde, as it were, and recover their primary or normal relation to the cerebral cavity. The attentive study of this singular cranium has enabled me to recognise the existence only of *three* cranial vertebra, essentially related to the three higher senses.

The orbito-sphenoids, or lower segments of the fronto-neural arch, are rarely developed in birds, as distinct elements: the bones indicated as such, by Mr. Owen, having no real or separate existence; that learned author regards them as the neurapophyses of the fronto-neural arch, and the frontal bones as the expanded and divided frontal spine. As growth advanced, the diploë of the coalesced frontals would begin to expand, and the obliteration of the sutures connecting them to the facial bones, would enable the increased development of pneumaticity to invade the frontal portions of the nasals and the median process of the premaxillary, so as to render them tumid, and tilt their anterior walls forward, producing the marked distinction between the cranial and mandibular segments of these elements.

The frontal protuberance culminates at a height of eleven lines above the cranio-facial line, and seven lines above the highest part of the parietal tract: it slopes rapidly downwards to the supra-orbital edge; its anterior border is undefined, where the frontals coalesce with the expanded cranial portions of the nasals and premaxillary; its posterior boundary follows nearly the posterior margin of the frontal, passing on each side outwards and forwards from the mesial line to a point on the supra-orbital border, midway between the notch and the post-orbital process. A broad shallow furrow traverses the median line, rendering it sub-bilobed; at the posterior extremity of the groove, where the coronal fontanelle existed, is a small foramen leading into a canal capable of admitting a fine bristle; it perforates directly the cranial parietes and opens internally towards the apex of the cerebellar fossa, transmitting a vein from the scalp to the posterior cerebellar sinus: the thickness of the cranium, here, amounts to eight lines and a half. Behind the foramen is a transversely oblong band, three lines and two-thirds wide, and one line long; defined posteriorly by an ungueal fissure-like groove, the angles of which extend outwards, curving backwards: its extremities are also defined by grooves. From the foramen a venous groove passes outwards, along the posterior border of the frontal protuberance, and externally curves slightly backwards to the aperture or notch on the supra-orbital plate, leading to the furrow on the roof of the orbit; about half an inch external to this edge, it is joined by a semi-circular groove which sweeps inwards, convex anteriorly, over the summit of the protuberance, and, bending backwards, reaches the median furrow; finally diverging from it, to meet the posterior groove at the outer and anterior angle of the band just mentioned. Where the semi-lunar grooves diverge from each other behind, a triangular, slightly elevated tract is left on the floor of the median furrow, with its base separated from the osseous band by a slight groove. From the convexity of the semi-lunar groove, two others pass forwards, on the left side, to a furrow, which appears to indicate the anterior edge of the combined frontals; in the centre it reaches half way up the frontal slope, its angle extends to the suture between the antorbital process and prefrontal on each side.

The precise limit of the subtriangular frontal plate of the nasal, is undefined; the external limb rises to a higher level than the internal, and is less abruptly bent on the body: on the left side, a groove curving outwards as it retrogrades, and continuous in front with the fissure between the prefrontal and ecto-nasal limb, indicates the outer margin of the body; the upper would form a segment of the frontal furrow; the internal is denoted by an interrupted fissure-like groove, which may be traced upwards from the linear impression separating the premaxillary median process, and the inner limb of the nasal, along the upper beam of the mandible; it passes inwards as it ascends, but appears not to have come in contact with its fellow behind, being separated in its whole extent by the termination of the premaxillary nasal process; which ascends to touch the frontals mesially, its apex having been probably inserted into the frontal suture. The lateral moieties of this extremity are also separated by a median fissure-like groove, which disappears as it ascends; the left one is more tumid than the right, and its anterior bullose extremity overhangs the cranio-facial line; on the right side, the pneumatic diploë does not cease so abruptly, and has a tendency to invade the median mandibular stem. This portion of the premaxillary measures six lines across its base, and appears to ascend ten lines and a half to its apex. The cranio-facial line is notched on each side, by the termination of the nasal fissure; the distinction between the mandibular and cranial segments of the nasal is least marked on the right side: the part of the latter, immediately behind the broad outer limb, is more inflated than the upper and inner angle; the expanded portion being defined by a semi-lunar groove. The triangular frontal aspect of the prefrontal, from the compression of the anterior part of the cranium, is directed very obliquely outwards; the anterior edge being much in advance of the posterior, which is anchylosed to the antorbital process; a deep fissure separates the anterior or inner margin from the ecto-nasal limb; the base forms the upper rounded border of the lacrymal groove, and terminates anteriorly in an obtuse projecting angle. This surface is perforated by numerous vascular apertures.

The sub-crescentic supra-orbital tract is rough, and perforated by periosteal vascular foramina; a series of larger size extend from the notch on the antorbital process to the supra-orbital foramen or notch, and indicate its inner boundary; hence the supra-orbital plate appears formed, as it were, by a separate ossification of the periosteum extending outwards to protect the eyeball. The space in front of the semi-circular venous grooves is also minutely punctate, the apertures becoming larger anteriorly. The tabula externa of the pneumatic diploë, on the frontal slope, is thinned and inflated opposite the individual cells; and some of these have opened out; these appearances indicate that the skull in question belonged to a domesticated individual.

The parieto-mastoid tract is gently arched transversely, but ascends rapidly in the antero-posterior diameter. It is narrow mesially, but extends laterally so as to occupy two-thirds of the convex external edge of the cranium, behind the supra-orbital notch. The rhomboidal digastric impression occupies the posterior angle of this tract, its transverse diameter is less than one-fourth of the breadth of the cranium; its posterior external angle corresponds to a slight groove on the mastoid process leading to the mastoid notch; its broadly rounded inner and posterior angle is separated from the supra-occipital ridge by the hinder horn of the parietal surface; a smooth narrow tract intervenes between its anterior margin and the temporal notch, and is continued into the smooth external surface of the mastoid process; the superior and anterior angle touches the pyriform muscular area which surrounds the crotophyte impression in front; the posterior border, as already mentioned, is separated from the occipital muscular surface by the smooth convex edge extending from the canalicular elevation to the mastoid notch. The crescentic crotophyte impression, forms a shelving entrance internally and anteriorly to the temporal notch; on the left side, a sharp ridge separates from the anterior limb of this surface, a triangular segment impressing the post-orbital process, with its base external. Surrounding the crotophyte impression in front and within, is a subpyriform excavation; its apex is truncated on a level with the posterior margin of the temporal notch, whilst its rounded extremity is separated by a narrow scabrous tract, four lines in breadth, from the orbital margin, and is so abruptly sunk beneath the level of the frontal protuberance, as to lodge the point of the finger. This impression doubtless gave origin to a cutaneous muscle, *dermo-mastoideus,* which is inserted into the integument of the posterior surface of the neck; it would erect the feathers on the head of the *Dodo,* and push forward the hood-like cutaneous ridge.

The anterior horn of the parietal tract sweeps round the dermo-mastoid impression to be continued into the rough supra-orbital space, bounding it externally. The parietal surface is variously marked by small vascular impressions, but is destitute of the foramina so abundant on the frontal slope.

The *anterior* and lesser aspect of the cranium presents the deep olfactory fossæ, separated mesially by the thin anterior prolongation of the interorbital septum, which rests below on the sphenoidal rostrum. The single small olfactory foramen, on each side, opens directly, with an inclination outwards, into the base of its respective fossa; at their exit from the apex of the cerebral cavity, they diverge slightly from each other, and are separated by an interval of about four lines, corresponding to the breadth of the interorbital septum.

Before describing more minutely the formation of the olfactory fossæ in the Dodo, it will be necessary to consider them first in other Pigeons, by which we shall alone gain a correct conception of several peculiarities in the cranium of this extinct form. In all Pigeons, as in many other birds, the anterior extremity of the vertical osseous plate, forming the interorbital septum, advances beyond the junction of the nasal with the frontal bones; and is completely covered by the former, which meet in the median line posteriorly, but are separated anteriorly by the extremity of the nasal process of the premaxillary; hence no part of it appears mesially, behind the premaxillary and between the nasals, as in the *Emeu* and other *Struthionidæ*. From each side of the expanded upper border of this advanced portion of the septum, a thin lamina passes horizontally outwards; contracting rapidly from before in the antero-posterior diameter, it bends downwards and inwards, arching over the foramen for the transmission of the olfactory and ophthalmic nerves and accompanying vessels, to meet and be continued for a greater or less extent along the outer border of a vertically transverse, subtriangular plate, projecting outwards from the interorbital septum: this last commences beneath the common aperture, and increases in breadth as it descends; by its anchylosis with the inferior extremity of the prefrontal, it forms the anterior wall of the orbit, separating it from the open olfactory cavity in front.

For reasons which cannot be discussed here, I regard the interorbital septum as the compressed body of the third and last, or most anterior of the cranial vertebræ; and the processes just mentioned, as ossified portions of the ethmoid or olfactory capsule; the superior I have hitherto denominated the turbinated, and the lower, the inferior ala of the ethmoid, and I shall continue to use these terms in the remainder of this description. By the sphenoidal rostrum, or rostrum simply, I understand the anterior prolongation of the sphenoid which supports the interorbital septum; it has been incorrectly considered as homologous with the anterior sphenoid in mammals, and hence has received the special appellation of presphenoid in Professor Owen's late paper on the Vertebrate skeleton;[1] whereas the interorbital septum in birds is the homologue of the mammalian presphenoid. The bone which has heretofore been denominated the lachrymal in birds, is undoubtedly the homologue of the prefrontal in the cranium of fishes and reptiles; the true lachrymal bone, which is external to the lachrymal duct, exists in certain Saurians, and in the *Crocodilidæ*; it does not occur in the higher Vertebrata, *Aves* and *Mammalia*, while the prefrontal only disappears in certain exceptional instances among mammals; in birds and mammals it has erroneously been regarded as the true lachrymal, and is so named even by the learned Hunterian Professor; this false homology masks one of the most beautiful instances of the unity of organization.

Having thus explained the meaning of the terms employed, we may return to our subject:—

The fissure remaining between the turbinated ala and the prefrontal, which in many birds transmits the upper diverticulum of the suborbital sinus, in several Pigeons, is diminished by the extension upwards of

[1] Reports of British Association, 1846.

the apex of the inferior ala, to join the interorbital septum, so as to form a bridge over the olfactory groove, behind that produced by the turbinated ala; the interval left between them, transmits a branch of the ophthalmic nerve with the accompanying vessels, which groove the outer surface of the turbinated ala, and escaping from between the nasal limbs are distributed to the nostrils. The outward expansion or development of the produced apex obliterates the fissure; the anterior wall of the orbit presenting only the olfactory outlet.

There is thus left a space between the turbinated ala and the prefrontal, which is closed behind by the outward extension of the former; it lodges a part of the subocular pneumatic sinus, from which the prefrontal receives air directly, by a large aperture on its inner surface. The compressed cavity internal to the turbinated ala is wider above and below, narrowest in the centre, where the olfactory orifice opens into it; the apposition of the pituitary membrane with that of the pneumatic sinus beneath the lachrymal duct bounds it externally, and below it is continued over the groove on the inferior ala to open into the posterior nares by the concavity of the nasal process of the palatine bone.

In *Goura*, the prefronto-ethmoidal fissure is not obliterated. In *Treron*, *Geophaps*, and *Calœnas*, it is completely closed; in *Carpophaga*, *Ptilinopus*, and *Didunculus*, only partially so.

In *Treron*, *Didunculus*, and *Calœnas*, &c., the turbinated ala is so curved outwards or evasated, as to come into contact with the apex of an inwardly inclined, subtriangular projection from the anterior margin of the prefrontal, supporting the termination of the lachrymal duct; and thus the pneumatic space is divided into two compartments; in *Treron*, from the great expansion of the diploë, it is much reduced in size.

In the Dodo, the prolongation of the interorbital septum, and the turbinated alæ, project about five lines beyond the junction of the cranium and mandibular apparatus; completely concealed from above by the latter, but not in contact with it, as in other Pigeons. The resilient hinge having retrograded to the cranio-facial line, space is left to permit of the downward flexion of the mandible; the remainder of this mechanism we shall see hereafter. The curved plate is much widened out to lodge the olfactory apparatus, and the convexity comes in contact with the prefrontal, in its whole length, at that part of the inner surface of the latter, which corresponds to the lachrymal groove externally; so that the subocular space is completely obliterated in the centre. The inferior ala is much compressed transversely and extended forward, so as to leave between it and the rostrum, a deep narrow groove; and the subocular space is reduced to a small irregular depression between its thin anterior edge and the prefrontal, with which it coalesces inferiorly. The prefronto-ethmoidal fissure is obliterated by the expansion of the posterior border of the turbinated ala, arching over the foramen that transmits the ophthalmic branch of the fifth nerve, which grooves the roof of the olfactory fossa. This aperture is diminished by an extension forwards of an osseous plate, from the interorbital septum outside of the foramen olfactorium; it forms the outer part of the floor of the olfactory fossa, and is, as it were, an ossification of the external wall of the periosteal tube, which conducts the olfactory peduncle to its exit at the antorbital foramen in most other birds; here, the tube in relation to the extremely short olfactory peduncle is much abbreviated, and its base widened out, serving to obliterate the space intervening between the antorbital and olfactory foramina. The olfactory fossa has a subhemispherical base, perforated by the single aperture for the transmission of the olfactory nerve; its floor presents the deep narrow groove just mentioned; the outer wall is perforated by the antorbital foramen about three lines anterior to the olfactory outlet. Each fossa is one inch two lines deep, and five lines wide at its anterior orifice; the height exclusive of the groove is six lines. The extremity of the high compressed rostrum is removed, exposing to view the very loose diploë enclosed by thin and elastic parietes; it probably terminated in a subacute apex. The anterior thickened margin of the inter-olfactory septum is concave anteriorly, and its lower portion ascends obliquely backwards, to a deep notch immediately below its upper end; it is sharp in the centre. The septum is translucent centrally, and its thinness gives increased space for the lodgement of the olfactory apparatus.

The relatively small *cerebral cavity* has its axis placed more horizontally, than in other Pigeons; so that the brain is, as it were, rotated on its transverse axis, and this rotation gives rise, or is related, to the verticality of the occipital facet.

The apex of the cerebral case is so depressed as to be nearly equidistant between the upper and under surfaces of the cranium, and to correspond externally to a point a little behind the centre of the groove on the interorbital septum. The frontals attain a median thickness of one inch and two or three lines, above the truncated apex, formed by a broad septum separating the olfactory foramina; which open directly into the bases of their respective fossæ. This septum is not homologous with the *crista galli* of the ethmoid, but is a prolongation upwards of the thick interorbital septum, or body of the olfactory vertebra, to coalesce with the frontals mesially, and thus to divide the anterior orifice of the cerebral tube into two foramina for the transmission of the olfactory peduncles, and so far to close it: the non-existence of any vertebral segment anterior to the frontal, permits the olfactory capsules to converge towards the median line and to be separated only by a thin septum, the prolongation of the anterior centrum; they are thus most exposed to the inspiratory currents of air loaded with odorous particles; the optic and auditory capsules, on the contrary, are situated between two adjacent vertebræ and project laterally. The thickness of the interorbital septum, beneath the olfactory foramina, is one inch two lines, and diminishes one half at the optic outlets. The length of the cerebral cavity, measured from the upper or lower margin of the foramen magnum to the olfactory septum, is one inch nine lines and a half; its breadth between the foramina for the transmission of the ophthalmic branch of the fifth is one inch; the extreme width of the cerebral fossæ is perhaps one inch nine lines, and its greatest height from the floor of the optic groove, probably about ten lines.

The basilar fossa for the lodgement of the medulla oblongata is slightly concave transversely, and rises towards the posterior clinoid plate, which projects with a subconvex border over the pit containing the orifices of the carotic canals, at the posterior part of the shallow and broad sella turcica; this plate is traversed at its base by the canal for the abducens nerve. The extreme length of the basilar fossa, from its posterior angle on the upper surface of the occipital condyle to the clinoid process, is one inch and a third of a line; its transverse diameter is equal to that of the lower segment of the foramen magnum: it presents, posteriorly and laterally, the condyloid foramen; and in front of it, the large infundibular pneumogastric orifice overhung by the vestibular prominence, which projects into the area of the foramen magnum at the centre of its lateral margin; a narrow convex ridge separates this aperture from the meatus auditorius internus, which has a subacute anterior edge running backwards on the vestibular convexity, between it and the petrosal fossa. The cerebellar fossa is relatively narrow, its length from the upper margin of the foramen magnum to its apex, dividing the cerebral fossa posteriorly, is one inch; its surface is considerably depressed beneath the level of the cerebral fossæ; it neither presents the longitudinal venous groove, nor transverse furrows corresponding to the laminæ of the cerebellum; along its margin posteriorly is seen the lateral venous groove terminating in the lateral occipital foramen; its lower angle, immediately above the foramen magnum, is perforated by the mesial occipital aperture. The fossa for the optic lobe is relatively very shallow internally, and its edges undefined; at its apex, beneath the lateral venous groove, there is a tumid pneumatic projection about the size of a large pea, overhanging the petrosal excavation. The thin floor of the optic fossa is pierced by the foramen giving passage to the ophthalmic branch of the fifth pair; internal to which it is grooved by the fourth nerve, which perforates the thin plate forming the posterior border of the optic foramen; at its deepest part it presents the foramen ovale. The oculo-motor aperture opens within the posterior border of the optic foramen, beneath the orifice for the trochlearis nerve. The form of the cerebral fossæ indicates that the cerebral lobes were broad and rounded in front, and elevated above the level of the cerebellum; the mesial ridge dividing them is low and obtuse, and subsides anteriorly towards the inter-olfactory septum, which is slightly carinate vertically; the outer edge of the olfactory foramen is subacute. The broad platform, formed by the interorbital septum between the olfactory and optic outlets, is convex in both diameters; its posterior, thick and rounded border projects over the groove lodging the optic chiasma. The surface of the cerebral fossa is smooth, and presents no trace of division into compartments.

The strong, much compressed and elongated *upper mandible,* corresponding to the two anterior thirds of the cranium, may be regarded as forming a three-sided pyramid; whose base is bevelled off in a direction downwards and forwards; the feebly uncinated apex projects beneath the level of the narrow palatine facet; and a plane replaces posteriorly the edge to which the broad lateral surfaces incline above. The thin upper margin of the base forms the hinge for the movements of the mandibular apparatus; from its anterior and inferior angles pass backwards the palatine bones to meet on the sphenoidal rostrum, while the slender pterygoids form a counter arch, springing from the inner angle of the inferior articular surface of the tympanic on either side; its crown abuts against that of the palatine. The strong sigmoid zygoma ascends from the external and inferior angle of the tympanic, to the centre of the outer edge of the base. The length of the upper mandible, measured from the cranio-facial line, is five inches eight lines, from the same line obliquely to the apex, five inches nine lines and a half; its greatest breadth is one inch seven lines; and its height, opposite the inferior angle, is one inch five lines and a half.

The high, compressed *core* occupies the apical third; it is formed by the premaxillary, whose strong mesial process, with the ento-nasal plate on each side, constitutes the upper beam of the mandible, a faint linear impression indicating their respective boundaries; the narrow lanceolate nasal fissure, perforating the basal two-thirds, divides the upper from the compressed lateral stems. The sutures between their elements are wholly obliterated; judging from analogy, the lateral process of the premaxillary bifurcates at a short distance behind the core, the inferior slip extending along the palatine surface, while the upper sinks into the outer aspect of the maxilla, and is wedged posteriorly between it and the expanded foot of the ecto-nasal limb, whose inferior boundary would probably be indicated by a line about an inch in length, drawn from the malar process to the lower border of the nasal fissure; along the posterior half of this line it meets the maxilla, which then passes internally to near the core, and forms the upper and lower, thick and rounded borders of the lateral beam, except for a short space anteriorly.

The lateral stem thus constituted has an elongated subtriangular form, with the truncated apex towards the core; the upper subconcave margin is three inches nine lines long; the base of two inches and three lines, ascends obliquely backwards, forming an angle of 125° with the palatine edge, which is of equal length. Viewed from above, the mandible presents a broad shallow excavation, impressing the outer surface of the lateral beam on each side, and extending from its prominent external basal edge, behind which the mandible is flatly compressed, forwards to the core.

This peculiarity in the skull of the Dodo, is due to the close approximation of the lateral stems in their anterior half; the mandible being most constricted about an inch behind the core, where its breadth is only seven lines, and its height one inch and three lines. The least breadth of each stem at the same place is three lines and one-third, and from this point they become rather broader as they proceed forwards, their lateral surfaces curving outwards to pass into those of the core; posteriorly they diverge and receive the broad termination of the mesial beam between their elongated upper angles. Their thickness also rapidly increases; that of the base is eight lines at the origin of the zygomatic process, but diminishes towards the angles, especially the superior. Viewed in front, the base of the core conceals the anterior portion of the floor of these fossæ, while behind, the posterior part appears shelving outwards; the upper angle is also seen to curve obliquely backwards and inwards, grooving the upper border of the stem. On tracing the groove downwards, it is seen to be continuous with a distinct broad impression, occupying the deepest part of the fossa, separated posteriorly from a narrow tract of the general excavation by a faint ridge, which curves downwards and forwards, becoming more distinct below; its concavity looks upwards and forwards, and its anterior angle corresponds to the most constricted part of the mandible. The anterior boundary is less distinct, it is formed by an oblique line descending forwards, in front of which the lateral stem increases in breadth. The chord of these fossæ, corresponding to a line drawn from the root of the zygoma to the greatest convexity of the core, is two inches and a half, and their greatest depth four lines and a half.

The lateral stems in front are nearly parallel and separated only by a narrow chink; at their base they slightly diverge, and the upper angles also are, as it were, twisted outward.

The base bounds in front the irregularly triangular lachrymal fissure; immediately below the centre of the external border is the origin of the zygoma, which passes backwards and downwards at an angle of 125° with the base: the lower half of the inner border is prolonged into a narrow, semi-lunar, antral plate, five lines in depth, its inner surface slopes obliquely outwards and is vertically convex; the outer is concave, a groove occupying its upper portion, at the superior angle of which is a large pneumatic aperture conveying air from the subocular sinus, and below it, several smaller ones; an elongated lacerated fissure opens into the *cancelli* of the expanded base beneath the root of the zygoma, along which its upper angle extends outwards. The lachrymal groove notches the base above the root of the zygoma and upper angle of the antral plate; it corresponds to that on the prefrontal behind. The pneumatic diploë terminates abruptly five lines and a half above this notch, fitting into the anterior extremity of the groove between the prefrontal and turbinated ala, while the projecting inner angle of the prefrontal is lodged in a slight depression on the outer surface of the protuberance. The prolonged upper angle formed by the ecto-nasal limb, is a thin elastic plate, five lines and a half long and about four lines broad, with a sharp external, and a thick rough internal, margin; it ascends to a higher level than the mesial beam, decussating it in the centre of its course; it expands slightly outwards above, and is separated by a narrow chink from the ento-nasal plate.

The mesial beam is formed anteriorly by the nasal process of the premaxillary; at its origin its basal area is subcircular; the rentrant angle between it and the lateral process on each side corresponds to an oblique fissure, leading forwards and outwards to the orifice of the horizontal vascular canal perforating the core; it tapers gradually backwards for the space of an inch, and then continues of nearly the same transverse diameter to the cranio-facial line, but diminishes rapidly in thickness; behind, it is supported by the ento-nasal plates which pass beneath it and meet each other in the mesial line inferiorly; an elongated narrow triangular portion appears externally on either side of the nasal process, and causes the mesial beam to increase rapidly in breadth in its posterior half. The length of the mesial beam is three inches nine lines; its least breadth towards its centre, where it is formed by the premaxillary nasal process is five lines and a half, its breadth at the fronto-facial line is one inch; and its transverse diameter, before expanding into the core, nine lines. The upper surface is flat behind with thick rounded borders; in front it declines gently outwards on each side from a faint mesial ridge, and the edge, descending as it advances, disappears from the expansion of the nasal process; the profile line has the same level for an inch posteriorly, and descends very gradually to the constricted part of the mandible, and then rises more rapidly to the same level, passing into the convex border of the core. The suture between the premaxillary and ento-nasal plate corresponds to a faint groove which passes forwards, and towards the centre of the stem attains its edge, thus defining the elongated triangular external segment of the ento-nasal plate, one inch eight lines long, and two lines and a half broad at its base; behind, it is continued upwards on the frontal slope, separating the inner margin of the nasal from the terminal extremity of the premaxillary nasal process; in front, it advances on the under surface of this process, diverging from its edge; it then bends more suddenly inwards for a short space, and about an inch from the distal extremity of the nasal process, retrogrades and speedily meets its fellow in the median line. The thin lancet-shaped extremities of the ento-nasal plates are thus defined, their apices being separated by the interposition of the nasal process; the ento-nasal plates continue to meet as far backwards as the free edge of the inter-olfactory septum; but where the mesial beam covers that septum and the turbinated alæ, they diverge to pass into the respective bodies of the nasals. This portion of the mesial beam is hence thinner and more flexible, and its upper surface, together with that of the adjacent ecto-nasal limbs, is excavated or thinned away in a semilunar tract, convex forwards, so as to give increased flexibility to the mandibular hinge; the greatest antero-posterior diameter of this tract is about eight lines, and it is minutely striated longitudinally.

The depressed posterior moiety of the nasal process is thus supported by the ento-nasal plates, which are concave laterally and meet below in a crest subsiding behind; so that the mesial beam is carinate inferiorly in its central moiety, and its section triangular; a groove furrows the keel posteriorly. The primitive division of the nasal process is also indicated by a very faint mesial groove, more perceptible posteriorly on the frontal slope, anteriorly it traverses the floor of the depression on the flattened posterior portion of the convex border of the core.

The wedge-shaped core supporting the short gnathotheca, is two inches long and about one inch four lines high in the centre; its greatest breadth is one inch one line. The lateral surfaces converge very gradually, are gently convex, and inferiorly towards the lower edge slightly impressed: a series of seven foramina occur on the right side, half an inch above the inferior border; the anterior is the largest, and forms the termination of the vascular canal, whose entrance is seen at the rentrant angle, between the nasal and maxillary processes; short divergent offsets from it open outwards, giving rise to the other foramina of that range; another set of four in number runs parallel to the upper border, the posterior is the largest, the anterior are narrow and slit-like, they are also the emergent orifices of vascular canals ascending from the primary one. Smaller foramina occur over the intervening space, which is also minutely grooved by the impressions of venous radicles. The upper border is gently convex, flattened and grooved behind; but sharper in front and prolonged into the feebly decurved apex, which is rounded off and not acuminated.

The palatine surface is concave, and bounded laterally by sharp alveolar edges, which are slightly involute; it is perforated by numerous large vascular apertures, and traversed by a mesial ridge; the palatine fissure grooves it posteriorly, widening out immediately before its termination to transmit the palatine nerves and vessels. The gently festooned alveolar edge is prolonged forwards into that of the apex, behind which it is concave inferiorly; it then descends towards the base of the core; and, lastly, rises into the line which separates the external surface of the lateral beam from the palatine tuberosity, which has its apex at the most constricted part of the mandible: this line ascends towards the root of the zygoma.

The base of the mandible and the lateral stem anteriorly present indications, in the opened-out osseous texture, of the domesticated condition in which the individual lived. The posterior angles of the external nasal fissures are nearly obliterated by the increasing breadth of the mesial beam as it retrogrades, and are closed by cellulo-fibrous tissue; the nostrils opening in front at the grooves formerly described.

The strong sigmoidal zygoma is formed by the *malar* and *zygomatic* styles, which coalesce at an early period; the distinction between the malar and maxillary bones is obliterated sooner; it descends obliquely backwards, as already mentioned, at an angle of 125°, with the posterior or basal edge of the maxilla, and attains the lower and outer angle of the tympanic, after a course of two inches seven lines and a half. From its origin, at the junction of the ecto-nasal limb with the maxilla, it is directed backwards and slightly outwards to the prefrontal; behind, it is strongly arched externally, beneath the orbit; in the downward flexion of the upper mandible, the hinder extremity of the anterior segment touches the prefrontal, which is flattened and granular at the point of contact; in the relaxed condition, it is separated by a chink, one line in breadth. The anterior portion is triangularly prismatic; the outer vertical surface is furrowed; the lower presents the prolongation of the upper angle of the maxillary pneumatic foramen into a deep groove; the upper is bevelled off inwards to the lower: all these surfaces are rough and striated. The long posterior *segment* is compressed vertically, and slightly contracted at each extremity; the upper smooth surface produced, as it were, by the flattening of the upper edge of the anterior portion, is convex, and directed downwards and outwards in its anterior moiety, but grooved longitudinally behind; the inner edge is smooth and rounded posteriorly, and flattened vertically opposite the prefrontal; the outer is thicker behind than in front, where it overhangs the inferior groove, it rises into the upper edge of the prismatic portion: the inferior surface slopes upwards and inwards, and is faintly furrowed lengthwise at each end. The posterior extremity presents a convex articular facet, directed inwards, and adapted to the pit on the lower and outer angle of the tympanic; a groove surrounds its neck for the attachment of the capsular ligament; the outer edge anterior to it, is covered by articular cartilage, on which the external mandibular ligament glides. The greatest breadth is two lines, and the depth one and a half.

The vertically spoon-shaped *palatine* bones, separated by a narrow chink anteriorly, arch outwards from each other behind, and finally approximate on the rostrum; they enclose between them the inferior nasal fissure, divided in the recent state by the membraneous septum, into the *choanæ*. Each palatine is formed of a scimitar-shaped sub-horizontal lamina (*crest*), with the cutting edge external, attached anteriorly to the maxillary, five lines in front of its angle; posteriorly towards the rapidly incurving point, the back is flattened into an oblong plate moulded to the rostrum, on which it glides; a triangular curved lamella (*nasal process*) rises from its inner concave edge into the lachrymal vacuity, while a similar plate (*palatine process*) descends to bound the inferior nasal aperture.

The crest is thin, flexible, and horizontal anteriorly, where it is adapted to the tuberosity of the maxilla; behind, it diminishes from without slightly in breadth, is thickened and twisted on its axis so as to shelve downwards; it also curves outwards, and lastly sweeps inwards, contracting, to be attached to the anterior moiety of the lower edge of the sphenoidal plate; the outer edge of the free portion is thus concave in front and convex behind; the inner is uniformly concave.

The nasal process forms an elongated triangular curved plate, with the apex in front; concave towards the nasal cavity, and inclining slightly outwards below towards its lower border, which is attached to the inner margin of the crest: it is bent rapidly inwards, to be attached by its posterior edge, in an oblique line directed downwards and backwards, to the anterior edge of the sphenoidal plate: its upper border, in its anterior moiety, is separated by a narrow fissure from the antrum; behind it is slightly emarginate on either side of a convex projection; this border gives attachment to the fibrous membrane of the subocular sinus, which stretches from the antrum to the inferior ala of the ethmoid; its concavity opens upwards in front of the olfactory fossa, and is prolonged downwards by the palatine plate.

The palatine process is a low, triangular, and slightly curved lamina; its anterior margin is convex, subsiding in front towards the maxillary, it is rounded off into the obtuse apex behind; the posterior is shorter, subconcave, and terminates at the anterior part of the sphenoidal plate; its inner surface is concave in the antero-posterior diameter, but subconvex vertically; a groove, corresponding to the crest externally, indicates the junction of its concavity with that of the nasal process. The narrow, elliptical, posterior nasal fissure is, anteriorly, prolonged into the slit between the lateral mandibular beams; in the dried state of the soft parts, the anterior angle of the posterior nares was two inches two lines and a half anterior to the point where the palatines meet on the rostrum.

An oblique line, directed forwards and inwards from the convexity of the crest to its inner margin anterior to the subsidence of the palatine process, defines the fossa between the crest and the palatine process, which gives attachment to the fleshy fibres of the *internal pterygoid*; the depressed tract on the anterior part of the crest, gives attachment to the tendon of that muscle; its fleshy fibres also arise from the fossa between the nasal process and crest, which is concave vertically and convex horizontally; it is prolonged posteriorly into a deep rough depression on the outer surface of the sphenoidal plate, at the bottom of which is the pneumatic aperture. The palatine bone is almost destitute of pneumaticity; the length of its free portion is two inches and a half.

The *pterygoid* bone, one inch two lines and a half long, is curved like the human clavicle, convex externally in front, and concave behind, and formed by a thin narrow band, twisted on itself and expanded at both extremities. The outer edge is thick and rounded behind; in front it becomes thinner, being flattened inwards, so as to encroach on the upper aspect, which thus exhibits an elongated triangular tract passing into the base of the inner extremity. The inner edge presents, on the left side, an angular projection in the centre, the rudiment of the sphenoidal articular surface; anterior to which it is flattened outwards, extending to the apex of the inner facet; the inferior surface is thus grooved in its two anterior thirds. On the upper aspect, a bidentate crest extends from the outer extremity obliquely inwards, and subsides towards the centre, bounding internally a lanceolate space. The inner extremity is triangular; the surface gliding on the sphenoid is slightly convex, and its anterior edge is united by ligament to the palatine bone; the compressed, oval posterior extremity is impressed by a narrow deep concavity for articulation with the tympanic. The pterygoid is destitute of pneumaticity.

The *tympanic* bone, viewed from behind, is X shaped; the lower segment being most extended transversely. The internal limb above is bent backwards, and supports an oblong articular tubercle, forming the posterior condyle, around its base are several foramina; the anterior is larger and covered by a triangular synovial surface, the apex behind its inner angle extends into a curved linear strip connecting them; the ligamentous groove is most distinct anteriorly and externally in both. A deep circular concavity, with a reticulate floor pierced by numerous pneumatic apertures, separates them posteriorly. A slight ridge, leading from the external condyle to an inflected notch at the centre of the outer concave edge, indicates the attachment of the membrana tympani. The inner margin is more defined, less curved, and exhibits a semi-obovate outline. Viewed externally, the figure is cruciform; the lower and outer angle projecting, and impressed by a deep pit for the extremity of the zygoma. The orbital process, corresponding to the upper and inner limb, is formed by a thin, curved, triangular plate; the apex is broadly truncated, inflated and slightly deflected outwards above; the outer surface is convex and pitted for the external pterygoid muscle, except for a narrow tract beneath the apex, extending backwards along its upper margin, which runs outwards to the anterior condyle; its inferior angle is bent inwards and expanded into a narrow pterygoid convexity, at the apex of which is a pneumatic foramen. The inferior moiety of the external aspect is smooth, polished, and convex across the projecting angle.

The inner surface presents a tripartite fossa, deepest inferiorly, rough throughout; and obliterated towards the extremity of its anterior angle by the expansion of the diploë of the orbital process, which is subtranslucent: it gives insertion to the *levator* muscle. A triangular flattened surface, its base extending between the condyles, separates the upper limb from a slight concavity, which impresses the external surface above, and indents the orbital process.

The mandibular extremity is compressed in the antero-posterior diameter; widest externally, and constricted in the centre. In its internal half, it forms a narrow transversely extended and downwardly projecting crest; the outer segment is flattened vertically into a quadrate plate, and twisted from before backwards on its axis. A broad groove directed obliquely forwards renders its inner moiety subconcave, and hollows out the inner tubercle externally. Articular cartilage covers the backwardly sloping inferior surface of the outer condyle, dips into the groove, and expands over the thick rounded edge of the inner one, which presents a large reticulated pneumatic foramen, and over its anterior surface, at the external angle of which it is narrowest. The greatest diagonal is one inch four lines and a half, and the least one inch three lines; the width of the upper extremity is eight lines and a half, and that of the lower one inch; it is three lines broad where most constricted. The orbital process is seven lines long; from its apex, which is three lines and a half wide, to the zygomatic angle, is one inch and three lines. The width of the outer mandibular condyle is five lines, of the inner two lines and a half; the length of the former is four lines; that of the latter, five lines and a half.

The rami of the *lower jaw,* six inches in length, measured along the lower border, are separated by an interval of two inches seven lines between their angles, but unite at a short, acute symphysis, which ascends at an angle of about 45° with the lower margin. Each ramus is thin and curved, the convexity mounting, external to the palatine wedge, into the angle between the inferior margin of the maxilla and the zygoma. The greatest height is at the centre, and amounts to one inch; it diminishes slightly towards each extremity. The upper edge is sigmoidal, convex behind, but concave in front; the profile line of the sharp upper edge of the core ascending very gradually as it advances, and then curving rapidly downwards to the mesial line of the symphysis, which is broadly emarginate anteriorly, concave above and subangularly rounded below; its convexity is adapted to the concavity of the edge of the upper core, while the decurved apex of the latter is fitted to its emargination. The lower concave edge is rendered slightly convex towards the centre of the posterior moiety, by the projection downwards of the dentary and angular elements. The dentary piece, exclusive of the core, is subequal in length to the posterior segment of the ramus; which is formed by the coalescence of the surangular, angular, and articular elements; the opercular remains distinct, perhaps, to a late period of life.

The dentary bifurcates posteriorly; the upper limb is slit vertically, to receive the surangular, whose free portion, unoverlapped by the long narrow inner dentary plate, advances halfway to the core; the suture between it and the short deeper outer lamina, is seen along the upper edge. The lower limb extends backwards, contracting, along the outer surface of the angular; whose thick lower border, thinning anteriorly, passes forwards, between it and the opercular, as far as the centre of the inferior edge of the ramus. The irregularly lanceolate opercular, partly covers the angular and dentary pieces, and rises to close the compressed space (*dental canal*) containing nerves and vessels; its posterior extremity extends along the inner surface of the angular, beneath the inflated portion of the articular; its superior border comes in contact with the inner and upper dentary lamina; anteriorly it terminates one inch behind the lower angle of the symphysis, its inferior edge diverging from that of the dentary; its length is about three inches, and its greatest depth six lines. The opercular begins to coalesce with the dentary, along the anterior part of its lower edge. A large vascular canal perforates the dentary very obliquely, running forward from the dental space; it opens externally nearer the upper than the lower border, about nine lines behind the symphysis, and passes into a deep groove, which, advancing towards the core, is concealed anteriorly by the perforated and undermined, posterior crenate edge of the latter, as it ascends obliquely backwards: its outer wall is pierced by two foramina, opening behind into a furrow, which disappears halfway to the dentary notch. The dental canal terminates, internally, above the anterior extremity of the opercular; the orifice is subdivided by a small process of the latter; a rough depressed triangular tract is continued forwards from it, to the thick posterior sub-vertical border of the symphysis, which is perforated horizontally by a vascular canal, emerging close to its lower edge, and midway between its angles; the mentary groove, already mentioned, joins this canal at its centre; externally, a series of five large apertures open upwards from it; three or four downwards, and one on the rough concave symphysial surface.

The fissure between the angular and articular pieces, is concealed by the dentary without, and the opercular within; that between the thin lamina of the surangular and the tumid rostrum of the articular element, is converted by the dentary fork into an elongated elliptical foramen, six lines and a half long, and one line and a half deep, its anterior angle corresponding to the posterior orifice of the dental space.

The posterior segment of the ramus increases in breadth as it retrogrades, and chiefly inwards in its hinder moiety, so as to present a large pyramidal internal angle; while the external is produced backwards into a compressed semi-lunar plate, with a thick and rough projecting edge, passing below into the smooth crest running upwards and inwards, between the basal facet, and the anterior aspect of the inner angle, which slopes obliquely forwards and outwards and passes into the inner surface of the ramus.

The basal or digastric facet is triangular, with the rough obtuse apex internal; the subconcave pitted surface passes externally into the inner convex aspect of the angular plate.

The complex articular surface is transversely oblong; without, it presents a longitudinal, slightly concave, reniform tract, with an external convex border; which plays on the outer flattened segment of the tympanic mandibular surface: within, it is deeply excavated for the reception of the ridge-like inner condyle; the concavity is directed inwards and forwards, becoming wider and shallower, its anterior depressed edge descending, while the posterior rises internally into a rough projection, rendering it concave transversely; this edge is narrow, rounded without, but internally, it presents the large oval pneumatic aperture. Synovial cartilage lines the inner half of its floor and its anterior surface, detaching a tract to line the reniform concavity.

The low short coronoid process is separated by an interval of six lines and a half from the articular surface, and corresponds to the junction of the four anterior fifths with the posterior; the upper edge of the surangular, anterior to it, is smooth and rounded.

An elongated tubercle extends downwards and forwards from the centre of the outer projecting border of the external segment of the articular surface, and its anterior angle is prolonged into a ridge, passing forwards to the extremity of the lower dentary limb. The outer surface of the posterior segment is thus divided diagonally, into two subequal surfaces; the posterior of which is triangular, most concave, and deeply pitted; it gives attachment to the muscles of the tongue, while the anterior furnishes insertion to the *M. temporalis*.

The pitted surface on the inner aspect of the jaw, for the attachment of the *M. pterygoideus internus* is defined anteriorly by an oblique irregular ridge, commencing at the upper edge a little anterior to the articular surface, and descending obliquely forwards to a groove, directed obliquely inwards and forwards, across the lower edge of the angular, from the extremity of the dentary to that of the opercular; it extends backwards on the anterior aspect of the inner angle.

The deeply concave pitted tract for the insertion of the *external pterygoid*, is anterior and superior to the former; it extends forwards, contracting to a groove, which leads to the posterior angle of the ramal vacuity; behind, it extends upwards to the coronoid edge.

Having thus given a detailed account of the skull of the Dodo, it now remains to contrast it generally with that of other Pigeons; for this purpose, and to supersede the necessity of lengthened descriptions, it has been thought desirable to give in Plate X., figures of some of the more remarkable and varied forms of the skull in that family, with a reduction of the head of the Dodo for comparison.

I am indebted to that eminent ornithologist, Sir W. Jardine, for permission to examine the bones remaining in the only specimen of the *Didunculus* in Europe; and thus I am enabled to confirm the opinion expressed by Mr. Gould, regarding the *columbine* affinities of that singular form, by the most certain of all tests, to wit, an examination of its osteological characters.

The most important and apparent difference between the skull in the Dodo and that in the lesser forms, depends on the small relative size of the brain and visual organs in the former, and the consequent abbreviation of the cranium and elongation of the basal part of the mandible. This difference, though readily explicable, might cause many, even, acute anatomists to overlook the family resemblances to that of Pigeons in the skull of this extinct bird. The happy appreciation of these by Reinhardt, entitles that learned zoologist to a high place among Palæontologists. In *Treron*, the extreme length of the cranial cavity is nine lines, and in *Goura*, one inch two lines; while in the Dodo, it is only one inch nine lines, little more than double the length of that cavity in the diminutive *Treron*. The outlet (foramen magnum) is also relatively less in the Dodo than in *Treron*. (Plate X., Fig. 2 *a*, 3 *a*.)

In the smaller Pigeons, the supra-occipital plate is less vertical and flattened than in the Dodo, being more arched transversely, and inclined obliquely backwards. The occipital condyle is less prominent, and the basilar protuberances for the insertion of the *M. recti capitis laterales majores* less apparent. The mesial supra-occipital aperture exists in all; and at an early period, it is very large and not separated by bone from the foramen magnum. Reference to the plates will indicate other minor differences, which may vary in the species of the same genus.

The superior facet in the smaller Pigeons is longer and narrower than in the Dodo, and the expansion of the frontal diploë less abrupt; the mastoidal angles, bearing the muscular impressions, are also bent downwards on the lateral aspects; while in the Dodo, they are thrown upwards by the great development of pneumaticity in the upper part of the lateral walls of the cranial cavity. The supra-orbital border is broadly concave and the interorbital region narrowed in ordinary Pigeons, and especially in *Geophaps*.

The profile is best seen in the respective figures; the dotted line in the Dodo indicates the probable outline before the development of the frontal protuberance. In *Treron* and *Goura*, there is a foramen as in the Dodo, perforating the cranium in the position of the coronal fontanelle; numerous venous grooves converge towards it. The form and relative proportion of the different muscular areas vary in the different genera; they are most deeply impressed in *Treron*.

In *Goura* there is a tendency to the development of the frontal protuberance opposite the apex of the cerebral cavity, and it is distinctly bilobed, the mesial line being traversed by a deep longitudinal furrow. The anterior portion of the coalesced frontals is elongated, as in other Pigeons, in relation to the great extent of the interorbital septum: the nasals touch each other in the median line, and their inner limbs are in contact with, and ultimately soldered to, the turbinated alæ. The hinge formed by the upper mandibular beam which here is chiefly constituted by the ento-nasal limbs, exists at the anterior edge of these plates, and in front of the posterior extremity of the nasal fissure. In the Dodo, on the contrary, this hinge is in the same transverse line as the hinder angle of the fissure, and the mesial beam is, as it were, started off the flat arch, formed by the turbinated alæ, which projects free under the mesial beam, separated from it by a space permitting the downward flexion of the mandible; the flexibility of the upper beam being increased by the thinning away of the part which conceals the free portion of the turbinated alæ. The bifid posterior extremity of the mesial process of the premaxillary is much smaller than in the Dodo, it passes beneath the coalesced nasals, resting on the upper edge of the inter-olfactory septum, and reaching about half-way to the frontal border of the nasal; in the Dodo, this extremity is much broader, and forms the principal part of the mesial beam at the hinge; and it reaches further back, separating the nasals mesially; its apex corresponding to the anterior extremity of the coronal suture. (See Plate X., Fig. 4 *b*).

In *Treron,* the diploë of the anterior portion of the coalesced frontals, is more expanded than in *Goura,* and the frontal aspect is convex transversely, and in the antero-posterior diameter; while in *Goura* it is concave transversely, and depressed longitudinally; the increased pneumaticity invades the nasals and overflows the extremity of the mesial beam, forming a tumid and abrupt cranio-facial line. The compact elastic extremity of the premaxillary process is wedged between this expansion and the inter-olfactory septum. The frontal aspect is depressed for a crescentic space, on each side, internal to the superciliary margin, and raised in the centre. (*Ib.* Fig. 3 *b.*)

In *Didunculus,* the forehead is flatter longitudinally than in *Treron,* but the broad extremity of the mesial mandibular beam is, in like manner, overhung by the tumid convex segment of the expanded and coalesced nasals; the central elevation of the frontal region is broader. (*Ib.* Fig. 1 *b*).

In all the lesser Pigeons, the arrangement of the mandibular hinge is essentially as in *Goura;* in *Goura, Geophaps,* and other slender-billed Pigeons, the ento-nasal limbs are very narrow posteriorly, hence the hinder angles of the nasal fissures are widened out, and expose to view the turbinated alæ. In *Calœnas,* the mesial beam is broader at the hinge, owing chiefly to the greater width of the nasal process of the premaxillary; and the posterior angles of the nasal fissures are reduced to narrow chinks, as in the Dodo; in *Treron* and *Didunculus,* these angles are also obliterated; but in all, the extremity of the nasal process of the premaxillary is concealed mesially by the junction of the nasals, and does not ascend on the frontal region to separate the nasal bones from each other, as in the Dodo.

The lateral aspect of the cranium in the lesser Pigeons differs from that in the Dodo, in the large relative size of the orbit, and in the great ratio which it bears to the temporal segment of the orbito-temporal fossa; the latter being diminished by the bending down of the mastoid element. The interorbital septum intervenes between the cerebral and olfactory fossæ: its junction with the coalesced frontals is traversed by the olfactory groove, which terminates in the antorbital foramen; the septum is thick and complete in *Treron,* in most other Pigeons it is thinner, and perforate in front of the common anterior boundary of the optic outlets: the floor of the cerebral cavity also is frequently membranous behind the olfactory foramen. In *Geophaps,* the post-orbital process is elongated, and nearly meets the post-temporal process of the mastoid; in *Didunculus,* the strong post-temporal plate is extended forwards and joins a slender bar from the post-orbital process, which completes externally the circular temporal outlet.

Inferiorly, the rostrum of the sphenoid in the lesser Pigeons is necessarily more elongated than in the Dodo; the pterygoid articular surfaces do not exist in the *Didunculus;* even in *Goura,* they are much reduced in size; in *Geophaps* and *Goura,* the groove on the rostrum leading from the common outlet of the Eustachian tubes is well marked, and the lateral venous depressions are also perceptible in *Goura;* the existence of these markings depending on the pneumatic expansion of the rostrum. The sphenoid and prefrontals are much inflated in *Treron* and *Geophaps.*

The ratio in length of the upper mandible to the cranium, in various forms of *Columbidæ,* is seen by reference to Plate X.; in the ordinary Pigeon they are subequal, but in the stronger-billed fruit-eating genera, the beak is shorter than the cranium, and in *Didunculus,* only half its length, while in the Dodo, it is twice as long. In the slender-billed species, the core is small, feebly hooked, and broadly rounded off apically; it is relatively large, broad, and depressed in *Geophaps;* in *Treron* it is stronger and wedge-shaped; but attains its maximum of development in *Didunculus,* where it is much compressed and more sharply uncinate than in the Dodo, assuming a pseudo-raptorial character, which, however, is negatived by the feeble osseous apex, and by the soft and foliated texture of the gnathotheca. The mesial beam is also much shortened in *Didunculus,* but its great breadth gives the necessary strength to the resilient hinge, required for the movements of this powerful beak; it is covered by a vestige of the cere, which is much extended in certain Trerons, but arrives at its greatest extension in the Dodo. The peculiar characters of the maxilla and the obliquity of the zygoma in Pigeons, have already been described; in *Didunculus,* the horizontal portion of the maxilla almost disappears, but the very strong basal segment ascends obliquely to join the broad ecto-nasal limb, and from their junction the zygoma descends to gain the tympanic. The palatine tuberosity or plane in other Pigeons, is, in *Didunculus,* replaced by the greatly developed funicular tendon of the *internal pterygoid* muscle, which arises from a strong tubercle at the base of the under surface of the core; as it passes backwards external to the palatine bone, it is covered within by the membrane of the subocular sinus, and below by that of the palate, forming a surface on which the convexity of the lower jaw glides. The short lunate nasal fissure in *Didunculus,* forms a striking contrast to its elongation in other Pigeons.

The shape of the palatine bone in the typical Pigeons, is well seen in *Treron,* (Plate X, Fig. 3 *c,*) and the deviations from it, in that of the Dodo, are readily accounted for, by the shortening of the sphenoid and the contraction of the mandible; the chief differences consisting, in the absence of the inflected portion of the palatine process, which in *Treron* diminishes the wide posterior nasal fissure; in the shortness of the sphenoidal plate, in relation to the abbreviated sphenoid; and in the less curvature of the nasal process, depending perhaps on the compression of the mandible.

In the *Didunculus,* the palate bone is much elongated, being attached anteriorly to the union between the very short lateral stem and the oblique ascending base of the maxilla, opposite the lower angle of the nasal fissure; the middle segment corresponding to the nasal process is drawn out, forming the extended base of the lachrymal vacuity; and from the great pneumaticity of the bone, the crest is expanded, narrowed, and subsides before reaching the sphenoidal plate. The nasal process is also but little apparent, the fossa between it and the crest being obliterated by the expansion of the diploë: the palatine process is a small curved triangular lamina, prolonging downwards the nasal concavity; it subsides behind at the anterior and inferior angle of the sphenoidal plate, and in front towards the termination of the crest; the large pneumatic aperture perforates the lower part of the sphenoidal plate. The small area afforded by the palatine, for the origin of the powerful *internal pterygoid,* is amply compensated by the great development of the tendon of that muscle.

The pterygoid bone is relatively longest in *Didunculus,* but has nearly the same form as in the Dodo, being destitute of pneumaticity, and of the sphenoidal articular surface; in *Geophaps* it has a similar shape, but articulates, as in most other Pigeons, with the sphenoid; it is much inflated in *Goura* &c., the pneumatic aperture being at the posterior extremity. The form of the inferior articular surface of the tympanic varies in the different genera of Pigeons; this surface in the Dodo closely resembles that in *Treron* and *Calœnas;* in *Geophaps,* the articular surface on the outer segment is much reduced in size, though that angle of the bone is much expanded, and chiefly backwards. In *Didunculus,* we observe the greatest deviation

2 D

from the typical form; the inner ridge-like condyle is extended in the antero-posterior diameter, and rests in a corresponding groove on the articular element of the lower jaw; while the flattened upper border of the thick and elevated outer wall of the latter, plays on a trochlear groove, extending externally along the posterior moiety of the base of the condyle. This ginglymoid joint permits the protrusion and retraction of the lower jaw, as in Parrots, for the purpose of unhusking fruit or seeds; the tympanic in the *Didunculus*, however, is readily distinguished from that in Parrots, by the double mastoid condyle.

The lower jaw varies in strength in the same ratio as the upper; it is more or less curved in the different genera, and to the greatest extent in the slender-billed species, in which the beak is arched downwards anteriorly. The dentary element is equal in length to the upper mandible; the posterior segment is much inflated in *Geophaps*; and in all has a triangular digastric facet, which in *Didunculus* slopes very obliquely forwards and inwards, but in *Geophaps* has nearly the same shape as in the Dodo; the external angular plate is not developed in the lesser Pigeons. The form of the articular surface necessarily varies with that of the coadapted aspect of the tympanic; the coronoid process is strongly developed in *Didunculus* and *Treron*. The vacuity between the angular, surangular, and dentary elements, is present in *Geophaps* and *Goura*, as in the Dodo; but is obliterated in *Didunculus* and *Treron*. The separation of the opercular element in the Dodo, indicates the incomplete development of the individual, and it occurs in the same condition in the specimen of *Geophaps* figured; but in the huge inert Dodo, it may remain unanchylosed longer than in the more active and volatile forms. The symphysis is broad and depressed in *Geophaps*, but is more acute and ascending in *Treron*, as in the Dodo.

In *Didunculus*, the dentary element is very strong, and the core is armed, on each side, with two small crenations, supporting corresponding teeth-like processes of the gnathotheca, as in the *Odontophorinæ* among the *Rasores*: and the symphysis is truncate anteriorly as in Parrots, the horny sheath covering the apex being abraded, in the specimen examined, so as to expose the cutis.

It is affirmed that this bird lives on bulbous roots; it may also live on hard-coated fruits and seeds, as suggested by Mr. Gould; the form of the articular surfaces of the tympanic and of the lower jaw, indicates the habitual employment of the lower mandible for decorticating roots, or unhusking fruits and seeds, after they have been crushed between the powerful jaws, the lower assisting especially by its dental armature; the depth of the impressions for the insertion of the masticatory muscles attests the strength of these actions.

The preceding details, accompanying the unrivalled lithographs of the skull of the Dodo (Plates VIII, IX, IX*), from the pencil of my esteemed friend Mr. Ford, (to whom I beg to return my sincere thanks,) will, I trust, be sufficient to remove any doubt regarding the *Columbine* affinities of that extinct form; the additional evidence furnished by the *foot* remains to be examined.

The evidence regarding the affinities of a newly discovered or extinct bird, deducible from the form and minute configuration of the metatarsus, is second in value only to that furnished by the skull.

The metatarsus, like the head, preserves, notwithstanding such variations as occur in the different genera and species of a common group, certain family characteristics, which are permanent; and which it is the province of the anatomist to eliminate, irrespective of absolute size.

The importance of this enquiry to the ornithologist, has led to its investigation in a general manner by Kessler,[1] whose researches will, I trust, be published by the Ray Society

[1] Bulletin de la Société Impériale des Naturalistes de Moscou. Année 1841.

The variation in shape of the small posterior or accessory metatarsus, which supports the hind toe, has hitherto been almost overlooked as a guide to classification, and farther observations are necessary to point out its real value; in the *Columbidæ*, the form of this bone is characteristic, and readily distinguishable from that of the corresponding element in the *Rasores*; although in some other respects, these orders closely approximate.

The number and relative length of the toes, the form and proportion of the constituent phalanges, and especially of the ungual segment, are also important elements for indicating the habits of birds, both as to progression and prehension. Although the *Columbidæ* are typically a perching group, still some of its members, as the Ground Pigeons (*Gourinæ*), seek their food chiefly, if not exclusively, on the ground, and require a corresponding adaptation in the form of the foot; which is not effected by a change in the shape of the metatarsi, or in the relative level of their trochlear extremities, in other words, by the assumption of the strictly ambulatory form of the foot, as in the *Rasores* and *Grallæ*; but chiefly by the abbreviation of the phalanges of the outer toe, which thus becomes shorter than the inner.

The same change takes place in the Dodo, which is a terrestrial representative of the *Treronine* group, just as the *Geophaps* is a less terrestrial member of the ordinary *Columbine* subtype.

The decayed and mutilated integuments were carefully removed from the remaining left foot of Tradescant's specimen by Dr. Kidd, the learned Professor of Anatomy and Medicine in the University of Oxford; and we are thus enabled to test the validity of the deduction arrived at from the study of the head, and vice versâ.

The opinion advanced by Professor Owen, after an examination of this interesting osseous relic, has been already mentioned; it is evidently based merely on the absolute size of the metatarsus, and the figures which he has furnished of its supposed affine, will serve for its refutation, while those given in Plate XI., will enable the reader to judge of the accuracy of Mr. Strickland's observations.

By authors, the principal metatarsus of birds is very generally termed the *tarso-metatarsus*, but improperly, as we have no evidence of the development at any period of the tarsal segment of the limb, or of its fusion with the three elements which coalesce to constitute the metatarsal bone; what has been regarded by some as the tarsal element, is simply the disjunct proximal epiphysis of the metatarsus.

The metatarsus of the Dodo (Plate XI, Fig. 1–6), which is five inches two lines and a half long, equals or exceeds in size that of the largest Raptorial bird, and is much greater than that of any of the known *Rasores*; in general form and proportions, it resembles most closely the corresponding bone in Pigeons, especially in the shorter-limbed arboreal species, as the *Treron*. The leading resemblances have already been stated,[1] and an examination of the figures (Plate XI.) will enable the general reader to verify them.

The great strength of this bone in Pigeons is remarkable, and the extended periphery is required to give an increase of surface, for the attachment of the powerful inter-osseous

[1] Part I. Chap. 1. p. 44.

muscles which move the toes to and from the axis of the foot; while the projection of the calcaneal process, which is supported by a highly developed buttress, gives much force to the action of the *flexor* muscles; hence the firmness of the grasp, so necessary in large-bodied birds with relatively small tarsi, is attained.

The posterior metatarsus in the Dodo (Plate XI, Fig. 7–10,) is nearly one-third of the length of the metatarsus, and measures one inch six lines; but the relative size of this bone is not greater than in any other known bird, for in *Lopholæmus* (ib. Fig. 43,) it is proportionally larger than in the Dodo.

The correspondence in the form and relative length of the anterior toes, and of their constituent phalanges in the Dodo, with those in the foot of *Geophaps*, one of the most terrestrial Pigeons, is well seen in Plate XII. The ungual phalanx (ib. Fig. 5, 5*a*,) forms a remarkable contrast in its shortness and blunted apex, and in the small size of the tubercle for the attachment of the flexor tendon, to the corresponding joint in the typical *Raptores*.

The supposed peculiarity in the Dodo, namely, "the equality of length of the metatarsus and proximal phalanx of the hind toe," is perhaps true as far as the *Columbidæ* are concerned; the difference however, if any, cannot be great in the Solitaire. The greater length of this phalanx in *Geophaps* and other terrestrial Pigeons, and the consequent elongation of the hind toe, is probably related to the persistent habit of rising occasionally from the ground and perching; while in the Dodo, which 'is not able to flie being so big,' the hind toe is much abbreviated and subservient only for support. The bluntness of the claws, and the shortness of the digits (Plates VI and XII), render it, at least, highly improbable that the Dodo could seize and hold *reptiles*, were such existing in its native isle; and the slowness of its pace would scarcely enable it to catch *littoral fishes* or *crustacea*, and in many parts of the coast these would be inaccessible to such heavy flightless birds, from the great and sudden rise of the shore above the water-edge.

Dimensions of the Metatarsi of the Dodo.

	inches.	lines.
METATARSUS.		
Length from the groove, on the middle trochlea, to the apex of the intercondyloid tubercle	5	$1\frac{1}{2}$
Least transverse diameter of the shaft		7
Antero-posterior diameter of the shaft opposite the articular facet for the posterior metatarsus		4
Greatest transverse diameter of the upper extremity	1	$5\frac{1}{2}$
Ditto antero-posterior of ditto, including ento-calcaneal process	1	$4\frac{1}{2}$
Projection of ento-calcaneal process		7
Width of the inferior extremity	1	$5\frac{1}{2}$
POSTERIOR METATARSUS.		
Length	1	$5\frac{1}{2}$
Breadth of the trochlea		6
Width of the lower extremity, including the styloid process		$9\frac{1}{3}$

The *shaft* of the metatarsus has nearly the same width (seven lines) in its middle third, but expands greatly towards each extremity and chiefly inwards, so that the inner margin is more concave than the external. The shaft is triangularly pyramidal in its two upper thirds, but compressed in the antero-posterior diameter below; the mesial ridge (calcaneal buttress), to which the lateral surfaces incline posteriorly, subsiding inferiorly towards the articular facet for the posterior metatarsus. The great development of the calcaneal buttress, which is based on the stem of the metatarsus, is one of the characteristics of that bone in Pigeons. The section of the shaft is hence triangular above, with the base in front, and the apex corresponding to the calcaneal ridge; below, the section is transversely oblong or suboval.

The anterior surface of the shaft is concave vertically in its upper and inner portion; in the rest of its extent it is straight longitudinally, and convex transversely. In its upper third, it presents a mesial elongated obovate concavity, between the prominent lateral metatarsal elements; the outer of which is most convex, and placed on a plane anterior to the inner, which is more expanded transversely, but less tumid. The median element forms the floor of the concavity, which is deepest beneath the overhanging edge of the proximal extremity; into it open the anterior orifices of the short canals, which perforate the bone in the antero-posterior diameter, and indicate, as in all other birds, its compound origin; both have a longitudinally oval form, but the internal is double the size of the external, the upper angle of which is partially concealed. The rough elongated and prominent oval tubercle which gives insertion to the tendon of the *M. tibialis anticus*, commences one line beneath the lower border of the inner foramen, and extends along the internal margin of the concavity, presenting a deep groove on its upper angle. Below, the anterior surface of the external metatarsal element slopes slightly backwards towards the broad outer border; while that of the inner elements is more rapidly rounded off centrally towards the inner edge.

The external third of the anterior surface, which twists on itself above, where it forms the outer wall of the concavity, is thus separated from the inner two-thirds, by a raised inter-muscular line, which descends from the inner margin of the external inter-osseous foramen to that of a well marked groove, commencing about half an inch above the oval aperture, or short oblique canal, that transmits, as usual, the tendon of the *M. adductor annularis*;[1] this muscle arises from the surface indicated. One line external to the inter-muscular ridge, a medullary foramen, directed downwards, perforates the shaft below its centre. Three distinct muscular impressions are met with internally; the outer descends from between the lower angles of the inter-osseous foramina, gradually increasing in breadth beneath the centre of the shaft, towards the middle trochlea, half an inch above which it terminates, and gives origin to the *Extensor medii*. The internal and inferior impression, from which the *M. adductor indicis* arises, extends from the *tibialis* tubercle as low as the external area, separated from it by an oblique sinuous line, which becomes fainter as it ascends, and disappears beyond the centre, being replaced by a slight groove; its upper and inner boundary descends obliquely inwards from the same tubercle, and reaches the inner margin towards its centre. The anterior surface of the inner metatarsal element, above this oblique line, gives attachment to the *M. extensor pollicis*; it is deeply pitted above on each side of a raised subcentral line, and below exhibits two or three faint grooves parallel to its lower boundary.

A small medullary foramen occurs nearly on a level with the lower angle of the posterior metatarsal facet, and one line external to the boundary between the surfaces for the *M. M. extensor medii* and *adductor indicis*, which are deeply pitted, especially below.

The external border, which is uncovered by muscle, is narrow below, but increases in breadth as it ascends, and turns round the convex outer metatarsal element so as to appear anteriorly; its anterior

[1] The inter-osseous muscles are named in relation to the median line of the body, not to the axis of the foot, as in English works on anatomy.

edge is faint, the posterior is sharper, and defines centrally the outer limit of the surface for the origin of the *Abductor annularis*; below, it becomes faint, and converges towards the anterior.

The internal border extends as a prominent narrow ridge along the inner element in its upper third; it then turns round towards the posterior aspect, subsiding from the antero-posterior expansion of the centre of the shaft, the anterior surface being broadly rounded off internally.

The triangular internal surface of the metatarsus has its base above, and extends below to the inner trochlea; in its upper third, it presents a deep pyramidal excavation for the origin of the *M. flexor brevis pollicis,* which also arises from the flat surface, extending beneath the fossa as far as the articular facet for the posterior metatarsus, situated at the junction of the middle and lower thirds of the shaft. This facet projects beyond the inner edge, and is covered by a transverse crescentic tract of synovial cartilage.

The external surface is flat, and separated from the inner by the rounded edge of the calcaneal buttress, which subsides towards the metatarsal articular facet. An intermuscular line descends from the posterior orifice of the external inter-osseous canal, and, becoming more prominent, sweeps outwards towards the external trochlea, half an inch above which it terminates; between this line, and the posterior edge of the broad external border, is an elongated tract of nearly uniform width, for the origin of the *M. abductor annularis.* The tendon of this muscle passes over a groove on the outer surface of the peduncle of the external trochlea, and is bound down by an oblique annular ligament, attached in front to a small oval tubercle on the outer edge, and behind to a rough ridge. A second more strongly marked inter-muscular line commences on the oblique outer aspect nine lines below the preceding, about two lines internal to it, and close to the posterior border of the subsiding calcaneal ridge, which it crosses as it runs obliquely inwards towards the articular facet, opposite which it becomes faint; about two lines lower down, it passes into a thick rough sigmoidal ridge, which terminates four lines above the inner trochlea, and gives attachment to the strong ligament connecting the metatarsi. An oblique sinus extending to the inner margin, lies between this ridge and the articular facet, and lodges the projecting lower angle of the upper extremity of the posterior metatarsus. Between this inter-muscular line and that previously mentioned, is a sub-triangular space which increases in breadth as it descends from the subsidence of the calcaneal buttress; the upper part lies on the outer surface, extending as high as the ecto-calcaneal process; the lower is deeply concave, and looks backwards; it gives origin to the *M. abductor indicis,*[1] the tendon of which is directed towards the inner intertrochlear notch, resting in a shallow groove, bounded internally by the outer edge of the elevated posterior surface of the peduncle of the inner trochlea, and externally by a rough elongated impression parallel to its inner boundary.

The *upper extremity,* viewed from above, presents in front the transversely reniform tibial articular surface, which is broad and rounded internally, narrower externally A semicircular non-articular tract lies behind in its concavity; the isthmus is raised into the prominent hemispherical intercondyloid tubercle, separating the inner large and deep subcircular condyloid fossa, from the shallow and smaller external one; whose anterior edge is bevelled downwards, and internally terminates in a slight pit at the base of the intercondyloid eminence, for the insertion of the external semilunar cartilage. The posterior and external subacute angle projects outwards, and its upper surface, which slopes backwards, gives attachment to the *M. peroneus brevis*; anterior to it, is a broad shallow notch lined by synovial membrane, immediately beneath which occurs the oval slightly elevated impression for the insertion of the external lateral ligament. The anterior angle presents a small transversely elongated tubercle beneath the edge of the condyloid fossa. The posterior and internal angle presents an oblong subvertical surface directed obliquely outwards and backwards; its lower and inner angle is tilted upwards, a sharp ridge descending from it on the floor of

[1] This muscle is not mentioned by Cuvier, Meckel, Owen, &c., although it probably exists in all birds.

the fossa for the *M. flexor brevis pollicis*; this surface gives attachment to the ligament of the great tibio-metatarsal sesamoid fibro-cartilage, which some have regarded, but improperly, as a tarsal element. Anterior to it is a narrow triangular tract, with the base above; the apex is roughened for the insertion of the internal lateral ligament, and is prolonged into the ridge-like inner border. Beneath the anterior and internal angle is the oblique oval prominent tubercle, for the insertion of a ligamentous band, binding down the tendon of the *M. extensor communis digitorum*; internally, it is attached to a short rough ridge, along the inner margin of the anterior concavity, separated from the preceding by a groove leading upwards and outwards, and impressing the anterior edge of the internal condyloid fossa, as it rises into the intercondyloid tubercle.

The smooth rounded upper border of the calcaneal ridge (*ento-calcaneal process*) projects backwards from the centre of the posterior border, and is raised in the middle into a slight convexity. The posterior angle of the ridge is expanded and flattened into an irregularly triangular rough plate (calcaneal tuberosity), with the base above, extending inwards and overhanging the internal muscular fossa, while the apex protrudes beyond the edge of the buttress; it gives attachment to the tendon of the *M. gastrocnemius*. Beneath the upper margin of the ento-calcaneal process, internally, is a curved oblong concavity, separated by a sharp ridge from the inner muscular fossa: it probably lodges an Haversian gland. The ecto-calcaneal process is formed by a short, thick plate (one line and two-thirds wide) projecting backwards, nearly midway between the internal process and the external edge; its free extremity is expanded and broadly grooved; from the inner lip of the furrow, an osseous bridge passes inwards to join the calcaneal ridge, converting the space between these processes into a canal (*calcaneal canal*), three lines in diameter above, and seven lines and a half long; below, it contracts slightly, and its inferior orifice is prolonged into a short triangular groove on the calcaneal ridge. This canal transmits the tendon of the *M. flexor perforans digitorum*. The deep groove posterior to it, between the calcaneal process and the outer lip of the ecto-calcaneal furrow, is closed in the recent state by a fibro-cartilaginous bridge; it gives passage to the tendon of the *M. flexor indicis perforatus* anteriorly, and to that of the *M. flexor indicis perforans et perforatus* posteriorly. The deep channel for the tendon of the *Flexor perforans pollicis*, furrows the outer surface of the ecto-calcaneal process, and is overhung by the projecting outer lip of its groove, from which a fibro-cartilaginous bridge extends, in the recent state, to a short faint ridge on the posterior surface of the external and outer angle of the proximal extremity; the groove is thus converted into a canal. External to the ridge just mentioned, is a short and shallow groove lined with synovial cartilage, over which plays the tendon of the *Peroneus medius* (Cuv.), as it proceeds to join the *perforated* tendon of the middle toe.

These grooves diminish in height from within outwards. In the recent state a broad shallow groove extends from the outer border of the calcaneal tuberosity to the slightly projecting edge of the external wall of the canal for the tendon of the *perforating flexor* of the hind toe; its floor is formed externally by the fibro-cartilaginous roof of that canal, centrally by the groove on the ecto-calcaneal process, and internally by the fibro-cartilaginous roof of the canal transmitting the *perforated* tendons of the inner toe; it is converted by the attachment of the fascia to its margins into a canal, which transmits most anteriorly the tendon of the *Flexor medii perforatus*, and posteriorly, that of the *Flexor medii perforans et perforatus*, and of the *Flexor annularis perforatus*, the former being internal, and the latter external.

Viewed from below, the *inferior extremity* presents the trochleæ arranged transversely, so as to form a small segment of a large circle; seen from before, the inferior surfaces of the two inner trochleæ lie in the same curve, while that of the outer is elevated four lines above the external margin of the middle one, and is nearly on the same transverse plane as the internal inter-trochlear notch. The elevation of the outer trochlea, and the abbreviation of the corresponding toe, renders the foot more adapted for progression. The internal trochlea, which is intermediate in size between the middle and external ones, the former being the larger, has its axis directed inwards, and is placed obliquely at the apex of a right triangular stem, which projects inwards, beyond a line drawn perpendicularly from the inner margin of the central part of the shaft; its inferior and internal margin is about a line above the plane of the middle trochlea; in front it is convex transversely, but its posterior and internal angle is elongated backwards and inwards, rendering it deeply concave behind. The middle trochlea is deeply grooved; the inner condyle is the most prominent anteriorly, but the external, below and behind; the groove expands, at its termination in front, into a sub-circular fossa impressing the stem. The outer trochlea anteriorly is more abruptly defined than the inner, and is slightly grooved; behind, the narrow outer condyle projects greatly. The sides of the trochlea are impressed with deep pits for the insertion of the strong lateral ligaments.

The metatarsus in the smaller Pigeons, and especially in the shorter limbed arboreal species as *Treron*, *Lopholæmus*, &c., has nearly the same form as that in the Dodo; but in many of the ground Pigeons (*Gourinæ*), it is relatively longer and more slender.

In *Treron* (Plate XI. Fig. 32–36), the inner metatarsal element is narrowed and flattened beneath the proximal extremity for the origin of the *M. extensor pollicis*, so as to look almost directly inwards; and the surface for the *M. adductor annularis* is relatively smaller, and also not visible from before. In *Lopholæmus* (*ib.* Fig. 38–42) and *Carpophaga*, the muscular surfaces are nearly as in the Dodo. The internal inter-osseous foramen is relatively larger, and the *tibialis* tubercle more remote from it than in the Dodo. In *Treron*, the trochleæ are nearly in the same curve, so also in *Lopholæmus*, and still more distinctly in *Carpophaga*; in all these, however, the inner trochlea is perceptibly more elevated than the outer. The outer edge is acute, forming a ridge separating the surfaces for the *M. M. adductor* and *abductor annularis*, and the areas which give origin to the *M. M. abductor annularis* and *abductor indicis*, are thus increased, especially that for the latter. In *Lopholæmus*, the large articular facet for the posterior metatarsal is placed nearly in the centre of the shaft; in *Treron* and *Carpophaga*, a little below it. In all the typical arboreal Pigeons, the ento-calcaneal process is elongated upwards at its expanded extremity; its upper edge is therefore concave (*ib.* Fig. 34, 40), not straight as in the Dodo; it also projects more than in the Dodo, and thus gives the *M. gastrocnemius* increased leverage. In *Lopholæmus* (*ib.* Fig. 41) and *Carpophaga*, the sculpturing of the ecto-calcaneal process is the same as in the Dodo; in *Treron* (*ib.* Fig. 35), the groove for the *perforated* tendons of the inner toe is converted into a canal. In *Lopholæmus*, the groove lodging the tendon of the *M. adductor annularis* is converted into a canal by an osseous bridge, leaving above it an aperture leading directly from the anterior to the posterior surface. In *Treron* and *Carpophaga*, the sharp posterior edge of the calcaneal buttress is slightly notched.

In *Columba* (Plate XII. Fig. 7), which represents a group intermediate, in habits and in the structure of the foot, between the arboreal and ground Pigeons, the form of the metatarsus so much resembles that in the Dodo, that it is difficult to specify the slight differences which exist. The outer border, uncovered by muscle, is broad like that in the Dodo, and twines round the outer metatarsal element, so as to appear on the anterior surface beneath the proximal extremity; this causes a diminution of the surfaces for the *M. M. abductor annularis* and *abductor indicis*, especially that for the latter. The upper border of the ento-calcaneal process is straight; the form of the ecto-calcaneal process is as in the Dodo, but the ridge separating the groove for the tendon of the *M. flexor perforans pollicis* from that for the tendon of the *Peroneus medius*, is more developed, and has a tendency to convert the former into a canal. The inner trochlea is less depressed than in the Dodo, the relative levels of the pullies being nearly as in *Treron*, &c.

This bone may be readily procured, for comparison with the figures of the metatarsus of the Dodo.

In *Didunculus* (Plate X. Fig. 9–9 *e*), and *Phaps* (Plate XI. Fig. 20–24), the metatarsus is elongated and slender, being of equal length in both, while they closely resemble each other in form and proportion; in each, the flattened outer border is still broader than in *Columba*, and the surfaces for the *M. M. abductor annularis* and *abductor indicis* are hence more reduced, from the encroachment of this border, which is widest in the centre, and contracts slightly downwards towards the inner trochlea. The area for the *M. extensor pollicis* is relatively larger than in the arboreal Pigeons. The upper border of the ento-calcaneal process is straight in both; in *Didunculus*, the groove for the tendon of the *M. flexor perforans pollicis* is converted into a canal (Plate X. Fig. 9 *d*), and that for the tendons of the *perforated flexors* of the inner toe, is also nearly closed. In *Phaps*, the outer ridge of the groove for the *M. flexor perforans pollicis* is very apparent, as in *Columba*. In *Didunculus*, the trochleæ are arranged exactly as in the Dodo; the groove for the tendon of the *M. adductor annularis*, is covered posteriorly by an osseous band, as in *Lopholæmus*, and in *Phaps*, where it is narrower. In *Phaps*, the inner trochlea is more elevated than in the Dodo, but the outer is more abbreviated than in *Didunculus*, and more like that in the Dodo; the posterior metatarsal facet in both, is placed below the junction of the lower with the upper two thirds of the bone.

The elongation and relative slenderness of the metatarsus, the great breadth and flatness of the outer border, and the position of the articular facet, are reproduced in the Solitaire; and the inner margin, which is acute in *Didunculus*, is replaced in *Phaps*, by a narrow plane as in the Solitaire.

In *Geophaps* (Plate XI. Fig. 26–30. Plate XII. Fig. 8), the metatarsus is shorter and more robust than in the two preceding species; and the outer margin, which is broad above, passes in the lower third of the shaft, into a narrow ridge separating the surfaces for the *M. M. adductor* and *abductor annularis*. The arrangement of the trochleæ is precisely the same as in the Dodo. In *Geophaps*, *Phaps*, and *Didunculus*, the *tibialis* tubercle encroaches on the inner inter-osseous foramen, as in the Solitaire, while in the Dodo, it is lower down. In *Geophaps*, as in *Didunculus*, the grooves for the *perforated* tendons of the inner toe, and the deep *flexor* tendon of the hind toe, are converted into canals.

In *Goura* (Plate XI. Fig. 11–15), the metatarsus has nearly the same form as in *Phaps*, but the outer border is relatively narrower. In *Phaps*, *Geophaps*, and *Goura*, the ecto-calcaneal, however, is thicker than in the Dodo, &c., and is grooved externally for the tendon of the deep *flexor* of the hind toe.

From these details we may therefore conclude, that the metatarsus of the Dodo possesses the family characters of that bone in the *Columbidæ*.

In the typical *Gallinæ*, the calcaneal buttress is feebly developed and speedily subsides, and the shaft is thus more compressed in the antero-posterior diameter; it is, however, as strongly marked as in Pigeons, in the short and robust prismatic metatarsus of *Pterocles*; and it is more apparent in the *Cracidæ* and *Megapodidæ* than in the common Cock. The external segment of the posterior surface is subconcave transversely, except in *Pterocles*. The ridge which supports the spur also distinguishes the metatarsus in the typical genera of the *Gallinæ*; that peculiar appendage is not the homologue of the *hallux*, as has often been supposed. Swainson long ago pointed out its true nature; it is really a portion of the dermo-skeleton, which becomes united to the metatarsal element of the ento-skeleton, by an extension of the ossific process in the intervening ligamentous texture; just as the teeth, which belong to the splanchnic division of the exo-skeleton, become anchylosed to the jaws in several fishes and reptiles. The *hind toe* is the true hallux, and is present in the great majority of birds. It has the normal number of phalanges, namely, two, and is supported by the accessory metatarsus; the outer, or fifth toe, is invariably absent in birds. In most of the *Gallinæ*, the tube which transmits the tendon of the *M. flexor digitorum perforans* pierces, as it were, the thickness of the ento-calcaneal process, and opens below upon, or to the inner side of the calcaneal buttress, which runs up to terminate in the thick peduncle forming the ecto-calcaneal process; in Pigeons and *Pterocles*, the ento-calcaneal process is thicker, and the tube terminates on the outer side of the buttress, which is continuous with that process. In *Pterocles* and in all the typical *Gallinæ*, the inner trochlea is somewhat more elevated than the external; but in the *Cracidæ*, it is more depressed. In the breadth of the inferior extremity, in the greater equality in size of the trochleæ, and in the great depression of the internal, beneath the level of the external one, the metatarsus of *Megapodius* approaches nearest to that in the terrestrial Pigeons. The variation in form and relative size of the surfaces giving attachment to the inter-osseous muscles, need not be dwelt upon here. The groove transmitting the tendon of the *M. adductor annularis* is in many of the *Gallinæ* converted into a canal by an osseous floor, as in certain Pigeons, both terrestrial and arboreal. The principal point, then, in which the metatarsus of the Dodo differs from that in the ordinary *Gallinæ*, is the greater development of the calcaneal buttress, which terminates superiorly in the ento-calcaneal process.

In the American Vulture (*Cathartes Californianus*), the metatarsus is compressed in the antero-posterior diameter, and deeply excavated in front, beneath the proximal extremity; the section of the shaft, both above and below, is therefore transversely oblong. The outer segment of the posterior surface is transversely subconcave; the calcaneal process is depressed, broad, and imperforate, presenting two broad shallow grooves posteriorly; from its centre extends downwards a moderately developed ridge subsiding towards the middle of the shaft; the external trochlea is more depressed than in the Dodo.

In the Vulture (*Gyps fulvus*) and Eagle (*Haliaëtus albicilla*), the shaft of the metatarsus is subtriedal, and the broad posterior surface presents a shallow groove in its whole extent. The wide, vertically concave, external border looks directly outwards in the Eagle, but in the Vulture has an inclination forwards and inwards; and the anterior surface slopes rapidly backwards towards the internal edge. The surfaces for the origin of the *M. M. adductor indicis* and *extensor medii* are very small, while that for the *Extensor pollicis* occupies the upper half of the anterior surface. The various grooves and tubes for the transmission of the *flexor* tendons in the Dodo and Pigeons, as well as in the *Gallinæ*, are represented by a single, deep, semilunar notch extending from the subquadrate plate, representing the ento-calcaneal process, to a prominent tubercle forming the outer and posterior angle of the proximal articular extremity; this process is not supported by a ridge descending on the posterior surface of the shaft, it is only a little more extended downwards basally, than at its free flattened extremity.

In the Vulture, the trochleæ are more nearly on a level; the external, as in *Cathartes*, being much lower than in the Dodo and terrestrial Pigeons. In the Eagle, the trochleæ are relatively narrower; the inner is placed on the same plane as the middle one, and its internal and posterior angle is more produced than in the Dodo and other Pigeons, or than in *Cathartes*. The inter-condyloid tubercle is relatively very small in the Vulture and Eagle. While the metatarsus in the Dodo is distinguished from that of the ordinary *Gallinæ*, only by a few and comparatively slight peculiarities, in which it approaches the typical Pigeons, it differs from the corresponding bone in the Vulture and Eagle, in the form of the shaft; in the presence of the complex ecto-calcaneal process, and of the highly developed calcaneal buttress; and in the greater elevation of the external trochlea: points of distinction so important as, even overlooking innumerable minor dissimilarities, to preclude any idea of the affinity of the Dodo to these raptorial forms. The metatarsus of the Dodo, in the presence of the calcaneal buttress, resembles that of *Cathartes* more than its homologue in the less aberrant raptorial birds; but this will not outweigh other important differences, and is such as not to indicate affinity, but simply a general resemblance in mechanical construction.

The *posterior metatarsus* in the Dodo (Plate XI. Fig. 7—10), is formed by a thick oblong plate, twisted on itself from behind forwards, and from within outwards; the line of flexure corresponding to its diagonal. The channel thus formed (*ib.* Fig. 9) lodges the *flexor* tendons of the hind toe. The lower extremity supports the transversely elongated trochlea, which is very slightly concave in front, but behind, it is grooved for the deep *flexor* tendon; it is broader internally than externally, and also projects backwards beyond the plane of the stem to a greater extent internally.

The thin outer margin of the triangular posterior portion of the stem is concave: its lower angle forms a subquadrate process projecting outwards beyond the trochlea (*styloid process*), which gives attachment to the annular ligament; its anterior surface is covered by a thin layer of synovial cartilage, the superficial *flexor* tendon of the inner toe gliding on it. The upper extremity, which articulates with the metatarsus, viewed from before, consists of a semicircular plate, forming the anterior wall of the channel; its thick convex, external border is roughened in front for the attachment of the strong inter-osseous ligament; the upper part of its subconcave anterior aspect is covered by articular cartilage, and the concavity probably gave origin to some fibres of the *M. adductor indicis.* The floor of the channel is wider below than above; and the lower untwisted portion of the stem projects obliquely backwards from the articular plate, which is perpendicular when in apposition with the metatarsus; the rounded inner margin of the former expands above into the triangular surface of the floor of the channel. The tendon of the *M. extensor pollicis* passes along the posterior surface, and is bound down immediately above the trochlea by an annular ligament, attached externally to a roughened portion of the outer edge, and internally to a narrow pit close to the inner border.

In all Pigeons, the shape of the posterior metatarsus is precisely the same as in the Dodo;—the styloid process exists in all (Plate XI. Fig. 16, 18, 25, 31, 37, 43). After observing that peculiar character, I was kindly allowed to test the *Columbine* affinity of the *Didunculus*, by removing its accessory metatarsal (Plate X. Fig. 10, 10 *a*), which proved to be a miniature of that in the Dodo. In the arboreal Pigeons, it is relatively larger than in the *Gourinæ*, and attains its maximum in *Lopholæmus.*

In the *Gallinæ*, the posterior metatarsus is relatively shorter, and the twist less distinctly marked than in the Dodo and Pigeons. In the common Cock, the curved plate is much thicker, and hence the channel more open, but the under or trochlear portion projects relatively farther back. In the *Cracidæ*, it is thinner, and more distinctly twisted; and from its greater elongation, its articular surface is placed lower down. In *Megapodius*, the outer margin of the curved plate is less concave, concealing from behind the expanded articular surface; but the essential distinction in all, lies in the absence of the styloid process.

In *Cathartes*, it is very small, subpyramidal, and not bent on itself; the anterior surface is nearly flat; its lower extremity is elevated considerably above the internal trochlea, so that the hind toe is, as in the typical *Gallinæ*, &c., above the plane of the heel.

In the Vulture, it is narrow transversely, and slightly twisted, the anterior surface being broadly concave, and the lower extremity placed nearly in contact with the inner margin of the metatarsus. In the Eagle,[1] also, the peculiar flexure of the posterior metatarsus almost disappears, the bone being nearly flat, and consequently it is readily distinguished from the corresponding bone in the Dodo; it is also destitute of the styloid process; and, as in *Cathartes* and the Vulture, the lower portion is shortened, and projects backwards only to a very slight extent. We therefore find, that as in the metatarsus, so also in the accessory metatarsal, the Dodo deviates more from the *Raptores* than from the *Gallinæ*, but is really distinct from both.

[1] The figure furnished by Mr. Owen, of this bone, *in situ*, is evidently taken from a badly mounted skeleton, in which it is placed in an unnatural position, as the ligaments would retain the accessory metatarsus close to the main one.

In the Dodo, the hind toe is about one third shorter than the inner, which, as in all strictly ground Pigeons, is distinctly longer than the outer; and the middle digit is not much longer than either of the lateral toes, but it is shorter than the metatarsus. The ungual joint of the outer toe only is preserved; the others are carefully restored in Plate XII. Fig. 1, 1 *a*, 1 *b*, as to length, from the foot covered with integuments, in the British Museum.[1] The phalanges have the usual form, and hence, it is unnecessary to enter into detail; the metatarsal articular surface of the first joint of each toe is seen in Plate XII. Fig. 2. The proximal (*ib.* Fig. 4,) and distal (*ib.* Fig. 4 *a,*) articular facets of one of the intermediate joints, viz., the second of the outer toe, are also figured.

	Hind Toe.		Inner.		Middle.		Outer.	
Extreme Length	inch.	lines.	inco.	lines.	inch.	lines.	inch.	lines.
of 1st Phalanx	1	4	1	6½	1	6	1	0½
2nd ditto.	.	..	..	11	1	..	..	7
3rd do.	..	..	.	..	..	9	..	6
4th do.	..	..	..	..	..	..	..	6
5th (ungual) do.	..	..	..	..	..	..	..	7

The proximal phalanx of the hind-toe is at least twice as long as the ungual segment; their combined length in the perfect foot in the British Museum is about two inches. It is longer than that of the outer, and shorter and flatter than that of the two inner digits, but equal in length to the posterior metatarsal. It may be distinguished from the other proximal phalanges, by the projection of the outer angle of its posterior extremity; by the shallow hinder articular concavity, and by the feeble development of its inter-condyloid ridge; by the great expansion of the distal extremity below, and the encroachment of the pits for the lateral ligaments on it above.

The proximal phalanx of the inner toe appears nearly double the length of the penultimate; its distal extremity is twisted slightly outwards towards the axis of the foot; the outer margin is also more concave than the inner; the concave metatarsal facet is reniform, the inner angle being most elongated; the absence of an inter-condyloid ridge on it, distinguishes this joint from the corresponding one of the middle toe, to which it is equal in length. The axis of the second phalanx is also directed outwards, but its distal extremity is bent inwards; it is strongly arched longitudinally, and its external margin is also more concave than the internal.

The proximal phalanx of the middle toe is broader and more robust than that of the inner, and is also twisted outwards towards its distal extremity, but its posterior articular surface is divided into two equal fossæ by an intermediate ridge, fitting into the groove on the middle trochlea. The distal extremity of the

[1] To Messrs. J. E. and G. R. Gray, I am under great obligations for the liberality with which they have allowed me to consult the public collections under their care. To the former, palæontologists and anatomists are most deeply indebted for the Osteological Collection now forming in the British Museum, which will enable the geologist to avail himself of the vast stores of fossil remains collected by the enlightened liberality of the Trustees. Hitherto, no means of turning them to account for the advancement of science have existed, as specimens cannot be removed for consultation, and few private persons possess collections, or can be at the expense of bringing skeletons to the British Museum to institute the necessary comparisons.

second phalanx is twisted inwards, as also that of the third, which is much arched longitudinally. These phalanges decrease distad, progressively, by one third.

The proximal phalanx of the outer toe is shorter than that of the other digits; its inner edge is more concave than the outer, the distal extremity being twisted inwards, while in the two inner it is bent outwards. The second, third, and fourth joints are much abbreviated, the two latter are nearly equal, and are only slightly shorter than the second, which is half the length of the first.

The ungual phalanx is a little longer than the second joint; it is short, curved, bluntly acuminate, and only slightly compressed laterally; the lateral surface presents a deep groove, the edges of which almost unite to form a canal towards the apex; the tubercle for the insertion of the *flexor* tendon is feebly developed; and the articular facet is equi-triangular, and slightly concave vertically.

In the arboreal Pigeons, as *Treron* (Plate XII. Fig. 6, 6 *a*), the inner toe is much shorter than the outer, and is nearly equalled in length by the hallux. The second and third joints of the outer toe are elongated; while in *Columba* (*ib.* Fig. 7, 7 *a*), they are shortened, and hence the lateral toes are nearly equal. In *Geophaps* (*ib.* Fig. 8, 8 *a*), these phalanges are more abbreviated, and the outer toe is shorter than the inner, as in the Dodo. All the ground Pigeons have this character more or less marked. The peripheral joints in these Pigeons, are relatively less abbreviated than in the Dodo; in it, the ambulatory modification of a strictly insessorial foot is carried to its maximum, but the persistence of typical characters is highly suggestive.

The arrangement of the tendons in the foot of the Dodo, is precisely the same as in that of Pigeons &c.; but throws no special light on its affinities. The sesamoid or glenoid fibro-cartilages on the plantar aspect of the metatarso-phalangeal articulations of the three anterior toes are represented as seen from above, in Plate XII. Fig. 2, *a*, *b*, *c*. They are firmly attached by ligament to the first phalanx, and but loosely to the peduncle of the trochlea by the reflected synovial membrane; anteriorly they are moulded to the trochlear surface, and posteriorly grooved for the *flexor* tendons; the theca converting the groove into a canal. The internal fibro-cartilage is acted on directly by the attachment of the *perforated flexor* tendon of the inner toe to the theca, just as the corresponding sesamoid fibro-cartilage at the tibio-metatarsal articulation is moved by the *Plantaris*: in the hind toe, the corresponding glenoid ligament was probably without a definite figure. The grooved posterior surface of the external glenoid ligament with the portion of the theca attached to it slit open, is seen at the top of fig. 3, Plate XII.

In addition to the strong lateral ligaments at the phalangeal articulations, there occurs a glenoid fibro-cartilage on the plantar aspect, which blends at the sides with the lateral ligaments. Like those above mentioned, each is firmly united to the distal phalanx, and moulded to the trochlear head of the proximal joint; and to it is attached directly the tendinous slip, which acts on the distal phalanx. At the last joint, they are feebly developed, and almost membranous; the deep flexor tendon being inserted into the tubercle beneath the articular facet of the ungual phalanx. In the Dodo, these fibro-cartilages remain, but are shrunk and indurated, and when covered by varnish, as in the Oxford foot, they very much resemble irregularly-shaped sesamoid bones. Those of the outer toe are shewn in fig. 3, *a* being the upper; the last is absent.

The celebrated physician and anatomist, Carus, when at Oxford, pointed out to Dr. Kidd a peculiar structure in the ossified tendons of the *flexor* muscles, and in his Travels[1] since published, he states "that the ossified tendons are divided into several pieces, connected by joints (internodia?), an arrangement

[1] "Ich machte Kidd auf eine besondre Bildung der verknöcherten Sehnen der Beugemuskeln aufmerksam."*—*England u. Schottland im Jahre*, 1844, vol. i. p. 375.

* "Die knöchernen Sehnen zerfallen in mehrere Stücken, welche durch Gelenke verbunden sind. Eine Einrichtung, die sonst an dergleichen Sehnenknochen mir nicht bekannt ist."

which he had not observed elsewhere in similar sesamoid bones (Sehnenknochen):" the supposed anomaly, however, disappears on moistening the foot, and examining the glenoid ligaments by transmitted light. A similar fibro-cartilaginous thickening of a crescentic form, with the concavity directed forwards, exists in the dorsal fibrous capsule of the joint connecting the two first phalanges of the middle toe, the tendon of the *Extensor communis digitorum* gliding over it.

The relative length of the toes, and of their individual segments, in the typical *Gallinæ*, are nearly as in the Dodo: the joints of each toe, exclusive of the ungual phalanges, decrease gradually in length distad; except in the outer, in which the penultimate is equal to, or longer than the second, and the second and third are occasionally equal. Like the metatarsus, the phalanges are relatively more robust in the Dodo. From the shortness of the accessory metatarsal bone, the hind toe is not on the same plane as the heel, when the digits are expanded and the foot in contact with a flat surface; but in the abberrant *Cracidæ* and *Megapodidæ*, it is more depressed.

In the Eagle, the hind toe is a little longer than the inner, and the latter is shorter but more robust than the outer, the middle being the longest, but slender when compared with the inner. In the hind toe, the ungual is equal to the proximal joint, which is stronger, broader, especially posteriorly, and longer in relation to the metatarsi than in the Dodo. The short, cuboidal proximal phalanx of the inner toe is only one third of the penultimate, and is sometimes anchylosed to it; the latter is nearly equal to the greatly developed ungual joint. The second joint of the middle digit is only one half of the length of the others, which are subequal; while in the outer, the penultimate is longer than the proximal, the intermediate joints are equal, and only half as long as the latter, the ungual phalanx being the longest. Thus in the two inner toes the ante-penultimate segments are much abbreviated, and in the outer, the two distal segments are relatively more elongated, but the three proximal, though shortened, have the same ratio to each other as in the Dodo, &c. The ungual phalanges progressively decrease in length and strength from within outwards, the hinder being the largest; the laterally compressed, subangular core is much curved and sharply uncinate; the vascular grooves in that of the Dodo are absent; the articular surface is more elongated and concave vertically, and the inferior tubercle is much larger.

In the Vulture, the middle toe much exceeds in length the lateral digits, which are nearly equal, and the hallux is shorter than the inner toe. The phalanges of the hind toe are equal; but the proximal joint of the inner toe is relatively twice as long as in the Eagle, but still only half the length of the distal phalanges which are subequal; in the middle toe the joints decrease in length, progressively, to the ungual, which, however, is longer than the penultimate phalanx; of the outer, the penultimate is shorter than the proximal phalanx, which is equal to the ungual; the second and third joints are also equal, and each only half as long as the penultimate. The Vulture thus exhibits a less raptorial foot than the Eagle. In *Cathartes*, the hallux is not half as long as the inner toe, which is shorter than the outer, and the middle digit is also much longer than the lateral toes; but the phalanges of the hind toe are equal. In the inner digit, the penultimate phalanx is shorter than the others, which are nearly equal; the joints of the middle toe decrease progressively to the ungual phalanx, which is longer than the penultimate; in the outer, the proximal is longer than the distal phalanx, the three intermediate being nearly equal, and about half as long as the first. The great strength of the claws is still remarkable in this modified raptorial sub-type.

The evidence furnished by the toes, corroborates that derived from a consideration of the metatarsi, regarding the non-raptorial affinities of the Dodo, and its closer approximation to the *Gallinæ*, from which, however, it is equally distinct.

CHAPTER II.

Osteology of the Solitaire.

(Plates XIII., XIV., and XV.)

THE osteological remains of the Solitaire, or supposed Dodo of Rodriguez, are few in number and imperfect, being either much mutilated, or thickly incrusted with stalagmite; sufficient, however, exists to indicate with certainty the true affinities of that extinct bird.

The particulars regarding the discovery of these bones, the probable localities in which they were found, and the principal inferences derived from the study of them, have already been fully described in this work (p. 46, *supra*).

Dimensions of the Cranium of the Solitaire.

	inches.	lines.
1. Length from the occipital condyle to the extremity of the inter-olfactory septum	3	5
2. Greatest breadth in front of the post-orbital processes	2	11
3. Height in the centre	1	8
4. From the occipital facet to the cranio-facial line	2	11
5. —— do. do. to the anterior margin of the orbit	2	7
6. —— the occipital condyle to the optic foramen	1	6
7. —— the optic foramen to the anterior margin of the orbit	1	1
8. —— the anterior margin of the temporal notch to that of the orbit . .	1	1

No allowance is made for the thickness of the incrustation, so that two lines at least must be deducted from some of these measurements.

The interesting *cranial* fragment is figured in Plate XIII. (Fig. 1–4), from drawings kindly furnished to us by M. de Blainville, the distinguished successor of Cuvier. It is most complete on the right side, but the paroccipital process is mutilated; inferiorly, the anterior part of the rostrum and the adjacent part of the inter-olfactory septum is destroyed; on the left side the prefrontal is broken away, and the parietes of the cerebral cavity removed; from the posterior angle of this vacuity a fissure passes inwards through the temporal notch, and another transversely through the occipital facet; but the par-occipital process is more perfect than on the right side. The mandible has been detached at the cranio-facial line, exposing to view the projecting inter-olfactory septum, and the turbinated alæ of the ethmoid, together with the entrance to the olfactory fossæ, as would be the case under similar circumstances in the Dodo. The fragment is so thickly covered with stalagmite as to render a minute description impossible, the deposit is thickest on the anterior and posterior parts of the upper surface, and the central tract hence appears depressed; an examination of the interior of the cranium has, however, convinced us, that this appearance is not due to any depression of the cranial roof.

A careful examination and contrast of the figures of the cranium of the Solitaire with those of the corresponding part in the Dodo, will prove the family affinity of these extinct forms, as well as their specific distinctness.

The cranium in the Solitaire is narrower and longer than in the Dodo, and is entirely destitute of the peculiar frontal protuberance; the individual elements, also, are less ventricose: its greatest breadth, as in the Dodo, is a little in front of the post-orbital processes; it probably decreased in width more rapidly forwards to the cranio-facial line. The orbits are more excavated, and the inter-orbital septum thinner, as in the more common forms of the *Columbidæ*; the prefrontal, especially below, is much less tumid than in the Dodo, and the rostrum is narrower. It resembles the cranium of the Dodo, and differs from that of the other known Pigeons, in the position of the olfactory fossæ, which are placed immediately in front of the cerebral cavity; the olfactory foramen, on each side, opening directly into the base of its respective fossa. The anchylosis of the prefrontal with the other elements of the cranium, may be regarded as one of the best proofs of the family affinity of the Solitaire and Dodo.

The occipital facet is vertical as in the Dodo; there is a cæcal excavation of the calcareous incrustation above the foramen magnum; does this indicate the mesial supra-occipital orifice in the Dodo and other Pigeons? The minute configuration of this aspect, as far as can be judged, closely resembles that in the Dodo, and the same may be said of the lateral and inferior facets; but the posterior angles of the upper surface are bent downwards, so as to encroach on the temporal segments of the orbito-temporal fossæ; hence the temporal notches are less apparent when viewed from above, and the surfaces bearing the muscular impressions, slope more rapidly downwards than in the Dodo, but to a less extent than in the common Pigeons. The prefronto-ethmoidal fissure is not so completely obliterated as in the Dodo; and the evasation of the turbinated ala is less marked, and more resembling that in *Goura*. The profile line would sweep, gently convex, downwards from the vertex to the cranio-facial line. The cranial cavity in its form corresponds to that in the Dodo.

Although we have ventured to differ from the illustrious Cuvier, who regarded this cranium as belonging to a gallinaceous bird, we trust we shall be excused; since a careful comparison of it with the skull of the Dodo at Oxford, has left no doubt on our minds of its affinity with that bird, and consequently with the *Columbidæ*. Unfortunately no portion of the upper mandible is yet known, but we may conjecture that it was less robust and more depressed than in the Dodo, and that it was only a little longer than the cranium. Judging from the figure given by Leguat, the caruncular ridge forming the line of demarcation between the peculiar columbine cere and the feathered skin of the head, was placed at the proximal extremity of the beak, and not on the forehead as in the Dodo.

We may hence suppose that the Solitaire is less remote from the *Treroninæ* than the Dodo, with which, however, it is inseparably united in the family *Didinæ*; the absence of the frontal protuberance and the other dissimilarities previously mentioned, establish provisionally its generic distinction, and the discovery of the beak will settle this question.

Less satisfactory evidence is deducible from the mutilated *sternum* (Plate XIII. Fig. 5 & 6), which is similarly incrusted with stalagmite. It is most perfect on the left side, the left costal process remaining, with the costal margin; but the external lateral processes are removed, and probably, also, a considerable portion posteriorly, including the mesial emarginations; the origin of the keel is, however, preserved. The form of the manubrial process differs from that in the sternum of the *Gallinæ*, but resembles that in most Pigeons; on the contrary, the costal process is less horizontal than in Pigeons, and ascends obliquely forwards as in the *Gallinæ*. The articular surfaces are apparently three in number, with intervening cellular spaces, as in *Goura*, &c. The sternum appears to have had a well developed keel, but unfortunately, there is not enough preserved to indicate its size and form; on the purpose of the keel in this flightless bird, I may refer to Mr. Strickland's remarks.[1] The anterior edge of the keel is grooved, and deeply concave, as in the *Gallinæ*, but the anterior, probably, precurved angle is deficient; the deep pneumatic excavation is seen posteriorly, beneath the manubrial process; and there is a deep depression at the root of the costal process, anteriorly, as in *Goura*, &c. This sternal fragment is four inches and a half long, and four inches broad.

The absence of any trace of the mesial fissure in the fragment preserved, and the form of the manubrial process, distinguish this sternum from that of the typical *Gallinæ*. The great development of the costal process, and the small number of costal articular facets, indicate a greater difference between this sternum and that of the Eagle or Vulture, &c. The presence of the keel is a proof of its non-affinity to the *Struthionidæ*.

The left *humerus* (Plate XIV. Fig. 1 to 3) in the Parisian Museum is also incrusted, and cannot be satisfactorily compared with that in Pigeons, which is the less to be regretted, since this bone furnishes no distinctive character; it is sufficient, that there is nothing to prevent its being regarded, as belonging to a *Columbine* form. It is four inches eight lines long; the *pectoral* crest was broken off before the bone became incrusted; the large pneumatic depression does not necessarily imply the existence of pneumaticity, which was probably absent. The short obtuse process of the rudimentary metacarpal of the thumb, covered with horn, as in *Chauna*, &c., formed "the little round mass under the feathers, as big as a musket ball," which the Solitaire employed as a weapon; the length of the wing, as indicated by that of the humerus, would give sufficient leverage for this purpose. Although the wing was wholly inadequate for flight, it might assist this large bird in running.

Dimensions of the bones of the leg of the Solitaire in the Andersonian Museum.

	inches.	lines.
1. Right femur. . . . . (Plate XIV. Fig. 4–7.)		
Length from the inter-condyloid notch to the upper surface of the neck	5	2½
—— from the external condyle to the extremity of the great trochanter	5	9
Transverse diameter of the shaft	0	8
Antero-posterior diameter of ditto	0	6½
Transverse diameter of the superior extremity	1	6
———————— of the lower ditto	1	4⅔
2. Left femur, with the extremities mutilated		
Transverse diameter of the shaft	0	8
Antero-posterior diameter of ditto	0	6⅔
3. Length of fragment of the right femur, with the extremities mutilated	4	11
Transverse diameter of the shaft	0	9½
Antero-posterior diameter of ditto	0	7½

[1] Part I. Chap. 2. p. 54, supra.

4. Left tibia . . . . (Plate XV. Fig. 1–1*a*.)

Length from the inter-condyloid groove to the termination of the fibular ridge	6	6
Transverse diameter of the shaft	0	8
Antero-posterior diameter of ditto	0	7
Breadth of the lower extremity	1	3½
Antero-posterior diameter of ditto	1	3

5. Left metatarsus . . . . (Plate XV. Fig. 2–2*b*.)

Probable length from the lower border of the middle trochlea to the summit of the inter-condyloid tubercle	7	3
Transverse diameter of the shaft		7⅔
Antero-posterior diameter of ditto, at the upper border of the articular surface for posterior metatarsal	0	5½
Transverse diameter of the lower extremity	1	6½
Distance from the upper border of the articular facet for the posterior metatarsal to the internal inter-trochlear notch	1	8½

6. Right metatarsus . . . . (Plate XV. Fig. 4.)

Antero-posterior diameter of the proximal articular surface, (to the calneal canal.)	0	7

Bones of the Solitaire in the Parisian Collection.

1. Left femur . . . . (Plate XIV. Fig. 8 to 10.)

Length from the inter-condyloid notch to the upper surface of the neck	6	4
——— from upper edge of the trochanter major to the external condyle	7	4
Transverse diameter of the shaft	0	11
Antero-posterior diameter of ditto	0	9
Transverse diameter of the upper extremity	2	2
——————— of the lower ditto	2	0

Thickness of incrustation more than one line.

2. Right metatarsal bone . . . . (Plate XV. Fig. 3–3 *d*.)

Length from the middle trochlear groove to the inter-condyloid tubercle	7	1
——— from the external trochlea to the external condyloid fossa .	6	7
——— from the internal ditto to the internal ditto . .	6	8
Breadth of the upper extremity	1	6
Antero-posterior diameter of ditto	1	5
Breadth of the lower extremity	1	9
Projection of the ento-calcaneal process	0	7

Thickness of incrustation about one line.

As formerly mentioned, three *femora* are contained in the Andersonian Museum, two right and one left; and a left femur incrusted with stalagmite, in the Parisian Collection (Plate XIV. Fig. 8–10). The most perfect of the former is the right femur (*ib.* Fig. 4–7), which is nearly entire, and belongs to a young individual; it is destitute of the pneumatic foramen, as in all Pigeons, except *Goura.* The femur, in general, does not yield any very distinctive character, but that in question resembles very closely in all respects, the same bone in Pigeons; it is not arched forwards as in the typical *Raptores.* In a left femur of the same size, but much mutilated at the extremities, and belonging probably to an adult female, the compact parietes of the shaft were one eighth of an inch externally, and one line and a half internally; while the femur of the gigantic Crane, of larger size, but possessed of pneumaticity, is only half a line in thickness. The cancellated tissue extends into the medullary cavity for a short distance at each end, and chiefly along the inner wall inferiorly; and the medullary cavity is lined by a thin osseous lamina, with few and minute perforations. The third specimen, also of the left side, is equally imperfect, but is much larger than either of the former, and belongs evidently to an adult male; it is larger than the corresponding bone in any gallinaceous bird, but is exactly equal in size to the coated femur (*ib.* Fig. 8–10), when allowance is made for the thickness of the incrustation. There can be no reasonable doubt, that these bones appertain to one and the same species, the diversity in size being attributable to differences in age and sex.

The fragment of the left *tibia* (Plate XV. Fig. 1, 1 *a*) closely resembles the corresponding bone in *Goura*; and judging from analogy, the upper third is removed, so that its length, when perfect, was probably nine inches and a half; the thickness of the parietes of the shaft is about one line. Its distal articular surface corresponds in size to the proximal extremity of the perfect metatarsus (*ib.* Fig. 3, 3 *c*). Its length indicates a bird of great stature, and fully justifies Leguat's statement, that the Solitaire is taller than a Turkey. Like the humerus and femur, this bone furnishes, in general, but few characters of importance; the osseous bridge under which the tendon of the *M. extensor communis digitorum* passes, distinguishes it at once from the corresponding bone in the *Struthionidæ.* The styloid inferior extremity of the *fibula* was probably more elongated downwards than in *Goura.*

Fortunately we are enabled to compare that important bone—the *metatarsus,* with its homologue in the Dodo, and thus to test the evidence afforded by the *cranium.* The right metatarsus (*ib.* Fig. 3–3 *d*), in the Parisian Collection, is covered with stalagmite; nevertheless, it enables us to establish the family affinity of these extinct birds. It differs from that in the Dodo, in its greater length and relative slenderness, the ratio being as seven to five: but the metatarsus of the Solitaire resembles that of the Dodo, in the form of the shaft; in the projection of the ento-calcaneal process, and in the great development of its supporting buttress; in the form of the ecto-calcaneal process; in the calcaneal tube opening on the outer surface of the buttress; in the presence of the articular facet for the accessory metatarsus; in the expansion of the distal extremity, and in the relative levels of the trochleæ. We are enabled to state more distinctly the differences in minute configuration between this bone in the Dodo and Solitaire, by an examination of the left metatarsus in the Andersonian Museum (*ib.* Fig. 2–2 *b*), which exactly resembles the Parisian specimen in form and size; and although much mutilated at the extremities, it supplies information otherwise unattainable. The proximal extremity of a right metatarsus (*ib.* Fig. 4), found with the preceding, and belonging to an immature individual, points out distinctly the relation of the calcaneal tube to the buttress.

The *posterior metatarsal* must have had the same dimensions as that in the Dodo, and we may safely conjecture, that it possessed the characteristic form of that bone in the *Columbidæ.* The *toes* would, perhaps, be less robust, and more elongated than in the Dodo.

The metatarsus of the Solitaire differs from that of the Dodo, not only in the greater elongation and antero-posterior expansion of the central portion of the shaft, but also in the greater breadth and transverse flatness of the external border, or surface uncovered by muscle, which does not curve round the upper part of the tumid external metatarsal pillar, as in the Dodo, but encroaches on and flattens that element, so that this margin is concave vertically in the Solitaire; while in the Dodo, it is slightly convex in its upper moiety; its anterior and posterior edges are acute, but the anterior in the Dodo is rounded off.

The concavity beneath the proximal extremity is deeper, and its floor angular; the outer wall formed by the central and external elements is only slightly concave; but the inner pillar is more convex and tumid than in the Dodo. The rounded surface for the attachment of the *Tibialis anticus* is in contact with the internal inter-osseous foramen, and extends on both walls of the concavity, a deep groove bisecting it; the outer segment is slightly raised, the inner impressed. The groove for the tendon of the *M. adductor annularis* is shorter and less distinct; and the tendon is transmitted through a canal, formed posteriorly by an osseous band, connecting the adjacent posterior edges of the peduncles of the two external trochleæ, a small oval foramen remaining above it for the transmission of vessels, and in front, as in all birds, by the bridge connecting them anteriorly. The line of demarcation between the surfaces for the *M. M. extensor pollicis* and *adductor indicis*, is imperceptible, and the inner limit of that for the *M. extensor medii* meets the outer at the lower extremity of the median concavity.

The inner border instead of being thin and ridge-like in its upper third, as in the Dodo, is replaced by a flat plane with a sharp posterior edge, the anterior is rounded off in young individuals; this plane slopes very slightly outwards, and terminates below at a rough projection situated at the junction of the upper and middle thirds of the bone, corresponding to a minute one in the Dodo; beneath which, the tumid inner part of the anterior aspect is broadly rounded off towards the internal surface. These surfaces are separated by a prolongation of the posterior margin of the replacing plane, which descends to the metatarsal facet, describing a curve convex posteriorly; the anterior margin of the plane is prolonged down on the convexity of the anterior surface, at first parallel to, and afterwards converging to the posterior, meeting it a little above the articular facet. The inner edge beneath this facet is less concave, being thinner than in the Dodo, and more extended inwards. The medullary foramina have the same relative position with regard to the shaft as in the Dodo. The fossa for the *M. flexor brevis pollicis* is narrower, from the replacement of the inner edge, and more elongated, extending to within an inch of the metatarsal articular facet. The edge of the calcaneal buttress is probably more convex. From the flattening of the outer border and the less projection of the calcaneal buttress, the surface for the *Abductor indicis* is much narrower than in the Dodo, and passes more directly into the groove lodging the tendon, which, however, is deeper and more distinctly defined than in the Dodo. The faint ridge which bounds this impression internally, and gives attachment to the inter-muscular ligament from which the *Flexor brevis pollicis* arises, subsides before reaching the articular facet.

The greater elongation and antero-posterior expansion of the central part of the shaft, and the breadth of the outer and inner borders are the most characteristic and essential differences between the metatarsus of the Solitaire and that of the Dodo. In all other respects they agree very closely; the dimensions, even, of the trochleæ and of the upper extremity, and the absolute height of the posterior metatarsal articular facet are alike in both. Those points in which the metatarsus of the Solitaire differs from that in the Dodo, are, in some measure, repeated in *Phaps*, which has nearly the same relation to *Geophaps* that the Solitaire has to the Dodo, in the proportionate lengths of the metatarsi. The metatarsus examined exhibits marks of disease similar to those found in the bones of birds dying in menageries, the compact osseous tissue is opened out along the lower moiety of the outer border, and in a circular space one inch beneath the proximal extremity, so that the bone is more acted on by atmospheric agencies at these places; and a small piece of the lower node is removed (Plate XV. Fig. 2 *a*, 2 *b*). The orifices of the minute periosteal Haversian canals are more distinct than usual, and give the surface a granular aspect. The parietes of the shaft in the immature specimen (*ib.* Fig. 4) are nearly one line thick; the medullary canal is divided, as usual, into three compartments by two thin partitions, which diverge as they pass from the anterior to the posterior wall; the cancellated tissue extends farther towards the middle of the shaft, in the narrow lateral, than in the wide central division.

We have now ascertained that the *cranium* of the Solitaire resembles that of the Dodo in numerous important points, differing in such respects only, as would justify us in regarding these birds as specifically distinct. The *metatarsus*, also, is principally distinguished from that in the Dodo, by such variations in size and proportion as might occur in species of the same genus. But in a small family, the members of which are confined to distinct localities, we are warranted from analogy, in regarding each as forming the type of a genus. The marked dissimilarity in external form between the Dodo and Solitaire, and the position of the caruncular ridge in the latter, together with the shorter beak, fully justify the establishment of another genus (*Pezophaps*) in the *Didinæ*, to include this lost form. That the Dodo and Solitaire belong to the same extinct sub-family of the *Columbidæ*, characterized chiefly by the peculiar structure of the cranium and rudimentary condition of the wings, no one will, we trust, doubt, who has carefully and impartially examined the evidence; the discovery of the osseous remains of the other extinct birds, supposed to belong to this group, will enable us more strictly to define its boundaries, and its alliances with the other sub-families of the Order *Columbæ*. We regard the Dodo, and its affine the Solitaire, as terrestrial flightless modifications of the *Treronine* sub-type, but having no immediate affinity with the other ground Pigeons, as *Goura, Calœnas*, &c., which are more directly allied to the ordinary *Columbinæ*.

For the reception of that modification of the *Treronine* sub-type, represented by the *Didunculus*, we propose to establish the sub-family *Gnathodontinæ*, in the hope that other members of the group remain to be discovered in the Polynesian Islands. The *Gnathodontinæ* are connected to the *Treroninæ* by the sub-genus *Toria*, which differs from the typical *Treron* in the abbreviation of the mandibles, and in the pseudo-raptorial form of the upper gnathotheca. The *Didunculus* is essentially a perching bird, but terrestrial affines probably exist, or have become extinct like the Dodo and Solitaire. The Pigeons form a perfectly isolated group of birds, having no direct affinity either with the *Insessores* or *Gallinæ*. The rasorial genus *Pterocles* approaches the Pigeons in the structure of the cranium, and in the form of the metatarsus; but it is destitute of the peculiar *columbine* cere, and the hind toe, when present, is rudimentary and elevated. The *Gallinæ*, then, approach the Pigeons through *Pterocles*, but no fusion of these groups is thus effected. From other considerations, the Prince of Canino and Col. Sykes had, also, previously recognised the approximation of *Pterocles* to the *Columbidæ*. The peculiar cere, and the great development of the nasal scales, are characteristic of the *Columbidæ*, and probably have some relation to the mode in which the nestlings are nourished. A milky fluid, analogous to the lacteal secretion in *Mammalia*, is elaborated by the thickened mucous membrane of the crop of the parents, and poured into its cavity, where it mixes with the macerating ingesta, and the young of certain species thrust their beaks into the throat of the parent, to obtain the food thus provided.

Postscript to Part II.

When this work was on the eve of publication, we received the *Bulletin de la Classe phys. math. de l'Académie Imp. de St. Pétersbourg,* vol. vii. No. 3, containing an abstract of a paper by Professor J. F. Brandt, entitled "Untersuchungen über die Verwandtschaften, die systematische Stellung, die geographische Verbreitung und die Vertilgung des Dodo, nebst Bemerkungen über die eim Vaterlande des Dodo, oder auf den Nachbarinseln desselben früher vorhandenen grossen Wadvögel." This memoir, which was read Dec. 17, 1847, contains the author's views of the affinities of the Dodo, which, it will be seen, differ considerably from our own. He states that after a diligent comparison of a cast of the Copenhagen Dodo-head with the osteological series in the Petersburg Museum, he had arrived at the following conclusions:—

"1. The Dodo, taken strictly, in regard either to the anatomy, or to the outer form of the head and foot, was not a Raptorial Bird, not even an anomalous one, although the last opinion has been adopted by several modern English and French naturalists of high reputation.

"2. The great difference in the form of its skull and beak from those of the Ostrich, equally forbids us to include it, as was formerly done, in that family of birds, although it approached them in its short wings, the texture of its plumage, its strong and (in general form) not very dissimilar feet, and the mode of scutulation of the tarsi.

"3. Neither can the Dodo be included among the Gallinaceous birds, on account of the very important differences of its cranial structure, and other discrepancies of outward form; although the form of its tarsus and the organization of its toes come very near to those of many Gallinæ.

"4. The Dodo agrees in the form of the majority of its cranial bones, and even in the shape of the beak, with the prevailing type of the Pigeons, as I had perceived, in common with my colleague v. Hamel. in the summer of 1846. Yet, considering the different form of the frontal, vertical, and occipital facets of its cranium, and the different shape and size of the lachrymal bone,[1] the palate bone, upper mandible, and maxillary continuation of the nasal, as well as the diversity of the wings, toes, and plumage, I am unable to refer it to the Pigeons, either immediately, or even as an aberrant form.

"5. The Dodo, a bird provided with divided toes and cursorial feet, is best classed in the order of Waders, among which it appears, from its many peculiarities (most of which, however, are quite referable to forms in this order), to be an anomalous link connecting several groups, a link which, for the reasons above given, inclines towards the Ostriches, and especially also towards the Pigeons.

"*a.* In regard to the cranial structure it approaches, among the Waders, most nearly to the Plovers, a group which also points, the most clearly of all Waders, to the type of the Pigeon's skull.[2] It inclines, it is true,

[1] Prefrontal of this Treatise.

[2] "The typical and great similarity of the skull in the Pigeons and Plovers is placed in juxtaposition in my treatise on the Dodo. One may accordingly regard the Plovers as Pigeon-forms, developed among the Waders, and

in a few points, more directly to the Pigeons than the Plovers do, yet these points, taken strictly, are such as the Pigeons have in common, not, indeed, with the *Charadrii*, but wholly with the *Porphyrio*, as well as with other groups of Waders. Moreover the Dodo, as already shown, differs from the Pigeons in the form of several of the cranial bones,—differences, nearly all of which exist also in the *Charadrii*, and occur as points of connection with different Wading birds.

"*b*. The remarkable form of the frontal region of the Dodo's skull indicates a combination of the frontal structure in *Chauna*, *Grus pavonina*, *Chionis*, and *Scolopax rusticola*, since, in regard to outline, it resembles *Chauna*; in the arching of its lower part, *Chionis*; in its great amount of arching generally, it is like *Grus pavonina*; in the very broad superior extremities of the lachrymal bone, trending towards the forehead, it agrees with *Scolopax rusticola*.

"*c*. The form of the crown and occiput of the Dodo reminds us of *Porphyrio*, *Grus pavonina*, the *Gallinæ*, &c., but not of the Pigeons.

"*d*. The elevated upper mandible of the Dodo, in which it differs from the *Charadrii* and Pigeons, refers us to *Ciconia*, *Tantalus*, *Ibis*.

"*e*. The broad maxillary continuation of the nasal bone in the Dodo, points to *Ciconia* and *Porphyrio*.

"*f*. The palatines of the Dodo, which do not slope outwards at the inner margin of their anterior extremity, are formed as in the *Gruinæ*, *Scolopacinæ*, and *Charadriinæ*, but not as in the Pigeons.

"*g*. The bones of the feet and toes in the Dodo agree best with those of *Hæmatopus*, among the Wading Birds.

"*h*. The naked forehead, cheeks, and gular region refer to *Tantalus*, *Grus leucogeranus*, and so to *Ciconia*, *Mycteria*, and many *Gallinæ*, much more than to the Vultures, and not at all to the Pigeons.

"*i*. The beak of the Dodo, in its general form, may be as correctly regarded to be a slightly modified colossal beak of a *Charadrius*, as of a Pigeon. On the other hand, it seems inadmissible to connect this bird with the Vulture, as it differs greatly therefrom in its short hooked extremity, only slightly emarginate at the lower edge.

"*k*. The nostrils, placed far forwards, and resembling perpendicular fissures, show a resemblance with those of *Chionis*, in part also with those of many Pigeons, but hardly with those of many Vultures (*nicht aber blos mit denen mancher Geier*).

"The Dodo may also be placed before the Dove-like *Charadrii*, as an anomalous form and a peculiar group of Waders, so that its affinity to Cranes, Storks, Woodcocks, Ibises, and Water-hens may be indicated; as I have done in a special table, which exhibits the single families of the Pigeons, *Gallinæ*, Ostriches, and Waders, arranged according to their relations of affinity. In the same table, also, the connections of the Dodo to the Ostriches and Pigeons are shown by dotted lines."

In a note appended to this paper, Professor Brandt thus relates the progress of his researches :—

"In order to establish more exactly my past, present, and future, wholly independent, opinion, with reference to Messrs. Strickland's and Melville's researches on the Dodo, I beg to make the following observations. Already in May, 1846, when Dr. Hamel had laid before the Academy a cast of the Copenhagen Dodo's head (Bull. Phys. Math. vol. v. p. 314), I invited him to join me in comparing the cast with the skulls of other birds in the Museum of the Academy. It soon resulted that the Dodo was no Vulture, Ostrich, or Galline, but rather a Pigeon-like bird. I soon after briefly communicated this result to M. Lichtenstein, and requested him to make it known to the Berlin Academy or the Natural History Society. It

greatly allied in the structure of their beaks; a relation which was unobserved by Strickland and Melville, inasmuch as they pronounced the Dodo to be actually a Pigeon."

was only in the autumn of 1847, that I had an opportunity of following up the observations in question more accurately, but my continued researches arrived at the conclusion that the Dodo was better placed as a cursorial bird in the vicinity of the Plovers, which are very like the Pigeon in the form of their skulls; especially as many others of its characters were also noticed in various wading birds. This result was already arrived at, and communicated to several friends (v. Baer, Kutorga, v. Middenderf, &c.) before I learnt Mr. Strickland's opinion."

The preceding remarks on the affinities of the Dodo, by Professor Brandt, would scarcely require any comment, were it not for the distinguished reputation of the Author as a Zoologist. It will readily be granted, that with all the materials extant for the decision of this question, at our command, we have more ample means of instituting the requisite comparisons, than the learned Professor, who had only a rough cast of the imperfect head at Copenhagen. The superficial resemblances, in the contour of the skull, and in the covering of the upper mandible, between Pigeons and Plovers, have been long known to naturalists; and were thus indicated by Swainson, in 1836 (Classification of Birds, vol. 2. p. 175), when speaking of the Plovers:—"Their heads are thick, and their eyes large, dark, and placed far back in the head; the bill is short, with the basal half soft, but the outer half becomes abruptly thick; and is often obsoletely notched, so as closely to resemble that of the Pigeon family, which in the Rasorial circle, appears to represent the great order of Waders." We were well acquainted with these superficial analogies; but, both from actual observation of the marked and essential differences in the structure of the cranium and foot in Pigeons, from that of the corresponding parts in Plovers, and also from a more correct interpretation of external characters, which, if rightly understood, are as valuable as those furnished by anatomical investigation, we were led to reject the hypothesis of any *direct* affinity existing between these families. Professor Brandt seems in this instance to have mistaken analogy for affinity, and in his anxiety to discover a link connecting dissevered groups, has wandered from the true method of investigation. The figures here given of the skull, and of the metatarsi, and the accurate representations of the integuments of the head and foot, will now enable our continental brethren to make the necessary comparisons, and to decide this interesting question for themselves; and it only remains to call their attention to the observations on the family characters of the skull in Pigeons, p. 97, *supra*.

APPENDIX, A.

Literal Translations of the Latin, French, Dutch, and German passages relating to the Dodo, in Part I., Ch. I.

1. Page 9. "Insula dicta præterquam," &c.

This island, besides being very fertile in terrestrial products, feeds vast numbers of birds, such as Turtle-doves, which occur in such plenty, that three of our men sometimes captured 150 in half a day, and might easily have taken more by hand, or killed them with sticks, if we had not been overloaded with the burden of them. Grey Parrots are also common there, and other birds, besides a large kind, bigger than our swans, with large heads, half of which is covered with skin like a hood. These birds want wings, in place of which are three or four blackish feathers. The tail consists of a few slender, curved feathers, of a grey colour. We called them Walckvögel, for this reason, that the longer they were boiled, the tougher and more uneatable they became. Their stomachs, however, and breasts were well tasted and easy to masticate. Another reason for the name was that we had an abundance of Turtle-doves, of a much sweeter and more agreeable flavour.

2. Page 9. "Déclaration de ce qu'avons veu," &c.

Fig. 1. Are Tortoises which frequent the land, deprived of paddles for swimming, of such size that they load a man; they crawl very stiffly, and catch crawfish a foot in length, which they eat.

Fig. 2. Is a bird, called by us Walckvögel, the size of a Swan. The rump is round, covered with two or three curled feathers; they have no wings, but in place of them three or four black feathers. We took a number of these birds, together with Turtle-doves and other birds, which were captured by our companions when they first visited the country, in quest of a deep and potable river where the ships could lie in safety. They returned in great joy, distributing their game to each ship, and we sailed the next day for this harbour, supplying each ship with a pilot from among those who had been there before. We cooked this bird, which was so tough that we could not boil it sufficiently, but eat it half raw. As soon as we reached the harbour, the Admiral sent us with several men into the country to seek for inhabitants, but we found none, only Turtle-doves and other birds in great abundance, which we took and killed, for as there was no one to scare them, they had no fear of us, but kept their places and allowed us to kill them. In short, it is a country abounding in fish and birds, insomuch that it exceeded all the others visited during the voyage.

Fig. 3. A Date-tree, the leaves of which are so large that a man may shelter himself from the rain under one of them, and when one bores a hole in them and puts in a pipe, there issues wine like dry wine, of a mild and sweet flavour: but when one keeps it three or four days, it becomes sour. It is called Palm-wine.

Fig. 4. Is a bird which we called *Rabos Forcados*, on account of their tails which are shaped like sheers. They are very tame, and when their wings are stretched they are nearly a fathom in length. The beak is long, and the birds are nearly black, with white breasts. They catch and eat flying-fish, also the intestines of fish and birds, as we proved with those which we captured, for when we were dressing them, and threw away the entrails, they seized and devoured the entrails and bowels of their comrades. They were very tough when cooked.

Fig. 5. Is a bird which we called Indian Crow, more than twice as big as the Parroquets, of two or three colours.

2 K

Fig. 6. Is a wild tree, on which we placed (as a memorial in case that ships should arrive) a tablet adorned with the arms of Holland, Zealand, and Amsterdam, so that others arriving here, might see that the Dutch had been there.

Fig. 7. This is a Palm-tree. Many of these trees were felled by our companions, and they cut out the bud marked A, a good cure for pains in the limbs. It is two or three feet long, white within, and sweet; some ate as many as seven or eight of them.

Fig. 8. Is a Bat, with a head like a Marmot. They fly here in great numbers, and hang in flocks to the trees; they sometimes fight and bite each other.

Fig. 9. Here the smith set up a forge, and wrought his iron; he also repaired some of the iron-work of the ships.

Fig. 10. Are huts which we built there of trees and leaves, for those who aided the smith and cooper at their work; so that we might start at the first opportunity.

Fig. 11. Here our chaplain, Philippe Pierre Delphois, a sincere and plain-spoken man, preached a very severe sermon, without sparing any one, twice during our stay in the island. One half of the crew attended it before dinner, and the other after. Here was Laurent (a Madagascar man) baptized, along with one or two of our own men.

Fig. 12. Here we applied ourselves to fishing, and took an incredible quantity, to wit, two barrels and a half at one haul, all of different colours.

3. Page 11. "Eodem quoque loco," &c.

In the same island are found many birds twice the size of Swans. The men named them *Walchstocken* or *Walckuëgels*, the flesh of which was not ill adapted for food. But as the same place furnished an abundance of Pigeons and Parrots, which were fat and well flavoured, our crew, neglecting the larger birds, preferred the more delicate and tender kinds, by feeding on which they solaced themselves in their troubles.

4. Page 12. "Cap. IV. *Gallinaceus Gallus peregrinus*," &c.

A foreign kind of Cock.—Of those eight ships which sailed from Holland in April, 1598, five came in sight of a mountainous island for which they gladly steered. While staying in the island, they noticed various kinds of birds, and among them a very strange one, of which I saw a figure rudely drawn in a Journal of that voyage which they published after their return, and from which the figure at the head of this chapter is copied.

This foreign bird was as large or larger than a Swan, but very different in form: for its head was large, covered as though with a membrane resembling a hood; the beak too was not flat, but thick and oblong, of a yellowish colour next the head, with the extremity black, the upper mandible hooked and curved, and in the lower was a bluish spot between the yellow and the black. They said that it was covered with few and short feathers, and had no wings, but, in place of them, four or five longish black quills. The hind part of the body was very fat and thick, and in place of a tail were four or five crisp curled feathers of a grey colour. Its legs were thick rather than long, the upper part as far as the knee covered with black feathers, the lower part and the feet yellowish; the feet were divided into four toes, the three longer ones directed forwards, and the fourth, which was shorter, turned backwards, and all of them furnished with black claws. The sailors called this bird in their own tongue, *Walgh-vogel*, that is, disgusting bird, partly because after a long boiling its flesh did not become more tender, but remained hard and indigestible, (except the breast and stomach which they found of no despicable flavour,) partly because they could get plenty of Turtle-doves which they found more delicate and savoury: it is therefore no wonder that they despised this bird and said that they could readily dispense with it. They said further that in its stomach certain stones were found, two of which I saw in the house of that accomplished man, Christian Porretus; they were of different forms, one full and rounded, the other uneven and angular, the former an inch in length, which I have figured at the feet of the bird, the latter larger and heavier, and both of a greyish colour. It is probable that they were picked up by the bird on the sea-shore and then devoured; and not formed in its stomach.

5. Page 13. "Op het lant onthouden," &c.

In this country occur Tortoises, *Wallichvogels,* Flamingos, Geese, Ducks, Field-hens, large and small Indian Crows, Doves, some of which have red tails (by eating which many of the crew were made sick), grey and green Parrots with long tails, some of which were there caught.

6. Page 16. "Verumenimverò, concinnatâ," &c.

After I had written down the history of this bird as well as I could, I happened to see in the house of Peter Pawius, Professor of Medicine in the University of Leyden, a leg cut off at the knee, and recently brought from Mauritius. It was not very long, but rather exceeded four inches from the knee to the bend of the foot; its thickness, however, was great, being nearly four inches in circumference, and it was covered with numerous scales, which in front were wider and yellow, but smaller and dusky behind. The upper part of the toes was also furnished with single broad scales, while the lower part was wholly callous. The toes were rather short for so thick a leg; for the length of the largest or middle one was not much over two inches, while that of the next to it was barely two inches, of the hind one an inch and a half. The claws of all were thick, hard, black, less than an inch long, but the claw of the hind toe was longer than the rest, and exceeded an inch.

7. Page 17. "On y trouve encore," &c.
"Men vinter ooc sekeren," &c.

They find there certain birds which some name *Dodaersen,* and others *Dronten.* Those who first arrived here called them *Walgh-voghels,* because they were able to procure plenty of others which were better. They are as large as a Swan, with small grey feathers, without wings or tail, having on their sides only small winglets, and behind four or five feathers more prominent than the rest. They have large thick feet, with a large clumsy beak and eyes, and have commonly in the stomach a stone as large as the fist. They are tolerable eating, but the stomach is the best part.

8. Page 17. "Pendant tout le temps," &c.
"Alle den tijt dat hier lagen," &c.

All the while they were here, they lived on Tortoises, Dodos, Pigeons, Turtle-doves, grey Parrots and other game, which they caught by hand in the woods. The flesh of the Land Tortoises was very well tasted. They salted and smoked it, and found it very serviceable, as were the Dodos which they salted.

9. Page 18. "Es hat auch daselbst," &c.

There are also many Birds, as Turtle-doves, grey Parrots, *Rabos forcados,* Field-hens, Partridges, and other birds in size like Swans, with large heads. They have a skin like a monk's cowl on the head, and no wings, but in place of them about 5 or 6 yellow feathers; likewise in place of a tail are 4 or 5 curled feathers. In colour they are grey; men call them *Totersten* or *Walckvögel*; they occur there in great plenty, insomuch that the Dutch daily caught and eat many of them. For not only these, but in general all the birds there are so tame that they killed the Turtle-doves as well as the other wild Pigeons and Parrots with sticks, and caught them by hand. They also captured the *Totersten* or *Walckvögel* with their hands, but were obliged to take good care that these birds did not bite them on the arms or legs with their beaks, which are very strong, thick and hooked; for they are wont to bite desperately hard.

10. Page 22. "J'ay veu dans l'isle Maurice," &c.

I have seen in Mauritius birds bigger than a Swan,[1] without feathers on the body, which is covered with a black down; the hinder part is round, the rump adorned with curled feathers as many in number as the bird is years old. In place of wings they have feathers like these last, black and curved, without webs. They have no

[1] The figure of this bird is in the second navigation of the Dutch to the East Indies, in the 29th day of the year 1598. They call it "bird of disgust."

tongues, the beak is large, curving a little downwards; their legs are long, scaly, with only three toes on each foot. It has a cry like a gosling, and is by no means so savoury to eat as the Flamingos and Ducks of which we have just spoken. They only lay one egg which is white, the size of a halfpenny roll, by the side of which they place a white stone the size of a hen's egg. They lay on grass which they collect, and make their nests in the forests; if one kills the young one, a grey stone is found in the gizzard. We call them Oiseaux de Nazaret.[2] The fat is excellent to give ease to the muscles and nerves.

11. Page 24. "De Dronte aliis Dodaers," &c.

Of the Dronte or Dodaers. Among the islands of the East Indies is reckoned that which by some is called *Cerne*, and by our countrymen, Mauritius, most famous for its black ebony. In this island a bird of wonderful form, called *Dronte*, abounds. In size it is between an Ostrich and a Turkey, from which it partly differs in form and partly agrees, especially with the African Ostrich, if you regard the rump, the quills, and the plumage; so that it seems like a pygmy among them in respect of the shortness of its legs. The head is large, clumsy, covered with a membrane like a hood. The eyes are large and black; the neck curved, prominent, and fat; the beak remarkably long and strong, of a bluish white, except the ends, of which the lower is black, the upper yellowish, and both pointed and hooked. The gape is hideous, enormously wide, as though formed for gluttony. The body is fat, round, and clothed with grey feathers in the manner of Ostriches. On each side, in place of quills, it is furnished with small feathered wings, of a yellowish grey, and behind the rump, in place of tail, with five curved plumes of the same colour. The legs are yellow, thick, but very short; the toes are four, stout, long, scaly, and the claws strong and black. The bird is slow and stupid, easily taken by the hunters. Their flesh, especially that of the breast, is fat, eatable, and so abundant that three or four *Drontes* have sometimes sufficed to feed a hundred seamen. If not well boiled, or old, they are more difficult of digestion, and when salted, are stored among the ship's provisions.

Pebbles of various form and size, of a grey colour, are found in the stomach of these birds, not however formed there, as the vulgar and the sailors believe, but swallowed on the sea shore; as though by this proof also it appeared that these birds agree with the nature of the Ostrich, since they swallow all kinds of hard substances without digesting them.

11. Page 25. "Num. 5 ist ein kopff," &c.

No. 5 is the head of a foreign Bird which Clusius names *Gallus peregrinus*, Nierenberg *Cygnus cucullatus*, and the Dutch *Walghvogel*, from the disgust which they are said to have taken to its hard flesh. The Dutch seem to have first discovered this bird in the island of Mauritius; and it is stated to have no wings, but in place of them two winglets, like the Emeu and the Penguins.

[1] Perhaps this name has been given them from having been found in the isle of Nazareth, which is higher up than that of Mauritius, in 17° S.

APPENDIX, B.

BIBLIOGRAPHY OF THE *DIDINÆ*.

Works which I have personally consulted are marked * (H. E. S.)

I. THE DODO.

A.D

*1598. (*Walckvögel*)—NECK (Jacob Cornelius van). Le second Livre, Journal ou Comptoir contenant le vray Discours et Narration historique du voyage faict par les huict Navires d'Amsterdam au mois de Mars l'An 1598. fol. Amsterdam, 1601; 2nd ed. 1609.——(Dutch) Waerachtigh Verhael van de Schip-vaert op Oost-Indien ghedaen by de acht Schepen, onder den Heer Admirael *Jacob van Neck* en de Vice-Admiral *Wybrand van Warwijck* van Amsterdam gezeylt in den jare 1598. 4to. Amsterdam, 1601; 1648, p. 6; another ed. 4to. Amst. 1650, p. 6.——(German) by *L. Hulsius*, Nürnberg, 1602; Franckfort, 1605.——(Latin) *De Bry*, Indiæ Orientalis partes IV, V. fol. Franckfort, 1601.——(English) London, 1601.—*Prévost*, Histoire générale des Voyages, 4to. Rouen, 1725; vol. 8. p. 123.—*Clusius*, Exotica, lib. v. ch. 4. p. 99.

*1602. (*Wallichvogels*)—HEEMSKERK (Jacob van). Journal of *Reyer Cornelisz* in "Begin ende Voortgangh van de Vereenighde Nederlantsche Geoctroyeerde Oostindische Compagnie." 4to. 1646. s. l. vol. 1.

*1602. (*Dodaarsen* or *Dronten*)—WEST-ZANEN (Willem van). Derde voornaemste Zee-getogt (der verbondene vrye Nederlanderen) na de Oost-Indien, gedaan met de Achinsche en Moluksche Vloten, onder de Ammiralen *Jacob Heemskerk* en *Wolfert Harmansz*. In den Jare 1601, 1602, 1603. Getrocken Uyt de naarstige aanteekeningen van *Willem van West-Zanen*, Schipper op de Bruin-Vis, en met eenige noodige byvoegselen vermeerdert, door *H. Soete-Boom*. 4to. Amsterdam, 1648, p. 21.

*1605. (*Gallinaceus Gallus peregrinus*)—CLUSIUS (C.) Exoticorum libri decem. fol. Raphelengii, 1605; lib. v. ch. 4, p. 100.

*1606. (*Dodaersen* or *Dronten*)—MATELIEF (Cornelius). Voyage in "Begin ende Voortgangh van de Vereen. Nederl. Geoctr. Oostind. Compagnie," v. 2. p. 5.——(French) Recueil des Voiages qui ont servi à l'établissement et au progrès de la Compagnie des Indes Orientales, formée dans les Provinces Unies des Pais-bas. 5 vols. 12mo. Amsterdam, 1702–1706; v. 3. p. 214.

*1607. (*Dodaersen*)—HAGEN (Stephen van der). Voyage in the "Tweede Deel van het Begin ende Voortgangh der Vereen. Nederl. Geoctr. Oostind. Compagnie." p. 88.——(French) Recueil des Voiages de la Comp. des Indes Or. v. 3. p. 195, 199.—*Prévost*, Hist. gén. des Voyages. v. 5. p. 246—*Van Soldt's* Voyage.

*1611. (*Totersten*)—VERHUFFEN (P. W.) Eylffter Schiffart, ander Theil, oder kurzer Verfolg und Continuirung der Reyse so von den Holl- und Seeländern in die Ost Indien mit neun grossen und vier kleinen Schiffen vom 1607 biss in dass 1612 Jahr verrichtet worden. *L. Hulsius*. 4to. Franckfort, 1613.

*1617. (........) BROECKE (Pieter van den). XXV jaarige Reyse-Beschryving naer Africa en Oost-Indien.

8vo. Lewarden, 1771. "Begin ende Voortgangh der Vereen. Nederl. Geoctr. Oostind. Compagnie." vol. 2. no. xvi. p. 102. plate 7.—*Thevenot*, Relations de divers Voyages curieux. vol. 1, Voyage de *Bontekoe*, pl. p. 5.

*1627. (*Dodo*)—HERBERT (Sir Thomas). Relation of some yeares' Travaile, begunne Anno 1626, into Afrique and the Greater Asia, especially the territories of the Persian Monarchie, and some parts of the Orientall Indies and Iles adiacent. fol. London, 1634, p. 211.—(2nd edition.) Some yeares' Travels into divers parts of Asia and Afrique, describing especially the two famous empires, the Persian and Great Mogull. Revised and enlarged by the Author. fol. London, 1638, p. 347.—(3rd edition.) Some years Travels into divers parts of Africa and Asia the Great. fol. London, 1677, p. 382.

*1635. (*Cygnus cucullatus*)—NIEREMBERG (J. D.) Historia Naturæ, maximè peregrinæ, libris xvi. distincta. fol. Antwerpiæ, 1635, p. 231.

*1638. (*Oiseaux de Nazaret*)—CAUCHE (François). Relation du Voyage de *F. Cauche*, in "Relations véritables et curieuses de l'Isle de Madagascar." 4to. Paris, 1651.

*1638. (*Dodo*)—L'ESTRANGE (Sir Hamon). Notes on Brown's Vulgar Errors, British Museum MSS. Sloane, 1839. 5.—*Wilkin's* edition of *Brown's* Works, v. 1. p. 369; v. 2. p. 173.

*1656. (*Dodar*)—TRADESCANT (John). Museum Tradescantianum, or a Collection of Rarities preserved at South Lambeth, near London. 12mo. 1656, p. 4.

*1657. (*Cygnus cucullatus*)—JOHNSTON (Johannes). Historiæ naturalis de Avibus libri vi. fol. Amstelodami, 1657, p. 122. pl. 56.

*1658. (*Dronte* or *Dodaers*)—PISO (Gulielmus). Additions to "*Jacobi Bontii* Historiæ naturalis et medicæ Indiæ Orientalis libri sex," in "*Gulielmi Pisonis* Medici Amstelædamensis de Indiæ utriusque re naturali et medicâ libri quatuordecim." fol. Amstelædami, 1658; lib. v. ch. 17. p. 70.

*1663. (*Dronte* or *Dodaers*)—THEVENOT (Melchizedec). Relations de divers Voyages curieux qui n'ont point esté publiées. 2 vols. fol. Paris, 1663.—Nouvelle édition, 2 vols. fol. Paris, 1696. Voy. de *Bontekoe*. pl. pp. 1, 5.

*1665. (*Dodo*)—HUBERT alias FORGES (Robert). A Catalogue of part of those Rarities collected in thirty years time with a great deal of Pains and Industry, by one of his Majestie's sworn Servants, R. H. alias Forges, Gentleman. 12mo. London, n. d.—(2nd ed.) A catalogue of many natural rarities with great industry, cost and thirty years travel in foraign countries collected by Robert Hubert alias Forges, Gent., and sworn servant to his Majesty. And daily to be seen at the place formerly called the Music House near the west end of St. Paul's Church. 12mo. London, 1665, p. 11.

*1666. (*Gallus peregrinus*)—OLEARIUS (Adam). Die Gottorfische Kunstkammer. 4to. Schleswig, 1666; 1674, pl. 13. f. 5.

*1668. (*Dodo*)—CHARLETON (Gualterus). Onomasticon Zoicon, plerorumque Animalium differentias et nomina propria pluribus linguis exponens. 4to. London, 1668, p. 113.

*1676. (*Cygnus cucullatus*)—WILLUGHBY (Franciscus). Ornithologiæ Libri tres, in quibus Aves omnes hactenus cognitæ in methodum naturis suis convenientem describuntur. fol. London, 1676, p. 107. pl. 27.—Translated into English and enlarged by *John Ray*. fol. London, 1678, p. 153. pl. 27.

*1677. (*Dodo*)—CHARLETON (Gualterus). Exercitationes de differentiis et nominibus Animalium. fol. Oxford, 1677, p. 117.

*1681. (*Dodo*)—HARRY (Benj.) A coppey of Mr. *Benj. Harry's* Journall when he was cheif mate of the Shippe Berkley Castle, Capt. *Wm. Talbot* then commander, on a voyage to the Coste and Bay, 1679, which voyage they wintered at the Maurrisshes. Brit. Mus. Addit. MSS. 3668. 11. D.

*1681. (*Dodo*)—GREW (Nehemiah). Musæum Regalis Societatis; or a Catalogue and description of the natural and artificial Rarities belonging to the Royal Society. fol. London, 1681, p. 60.

*1684. (*Gallus gallinaceus peregrinus*)—LLHWYD (Edward). Catalogus Animalium quæ in Museo Ashmoleano conservantur. MS. in Ashmolean Museum. Lib. Dni. Principalis Coll. Ænei Nasi, No. 29.

1688. ( )—Lacroix ( ). Relation des Iles d'Afrique.

*1700. (*Dodo*)—Hyde (Thomas). Historia Religionis veterum Persarum eorumque Magorum. 4to. Oxon. 1700 p. 312. pl. 7.

*1704. (*Dronte*)—Nieuhoff (John). Voyages and Travels to the E. Indies in *Churchill's* Collection of Voyages and Travels. 4 vols. fol. London, 1704; v. 2. p. 354.

*1713. (*Cygnus cucullatus*)—Ray (John). Synopsis methodica Avium et Piscium; opus posthumum. 12mo. London, 1713, p. 37.

*1752. (*Raphus*)—Mœhring (P. H. G.) Avium Genera. 12mo. Bremæ, 1752, p. 58.

*1757. (*Dodo*)—Edwards (George). Gleanings of Natural History, exhibiting figures of Quadrupeds, Birds, Fishes, Insects, &c. 3 vols. 4to. London, 1755–1764, pl. 294.

*1758. (*Struthio cucullatus*)—Linnæus (Carolus). Systema Naturæ per Regna tria Naturæ, editio decima. 2 vols. 8vo. Holmiæ, 1758; vol. 1. p. 155.

*1760. (*Raphus*)—Brisson (M. J.) Ornithologia, sive Synopsis methodica sistens Avium divisionem in Ordines, Sectiones, Genera, Species, ipsarumque Varietates. 6 vols. 4to. Paris, 1760; vol. 5. p 15.

*1767. (*Cynge étranger*)—Salerne ( ). L'Histoire Naturelle éclaircie dans une de ses parties principales, l'Ornithologie. fol. Paris, 1767, p. 80.

*1767. (*Didus ineptus*)—Linnæus (Carolus). Systema Naturæ per regna tria Naturæ, secundum classes, ordines, genera, species, cum characteribus, differentiis, synonymis, locis. ed. 12. 3 vols. 8vo. Holmiæ, 1767; vol. 1. p. 267.

*1770. (*Dronte* and *Oiseau de Nazare*)—Buffon (G. L. Le Clerc de). Histoire naturelle des Oiseaux. 9 vols. 4to. Paris, 1770–1783; vol. 1. pp. 480, 485.—Ed. 2. 10 vols. fol. Paris, 1771–1786; vol. 2. pp. 73, 77. —Nouvelle édition par *C. S. Sonnini*, 28 vols. 8vo. Paris, 1801–1805; vol. 4. pp. 336, 343. pl. 33. f. 1.

1773. (*Dodo*)—Seligmann (J. M.) Sammlung seltener Vögel. 8 vols. fol. Nürnb. 1749–1773; v. 8. pl. 84.

(*Dronte*)—Bomare (J. C. V. de). Dictionnaire raisonné universel d'Histoire Naturelle. 5 vols. 8vo. Paris, 1765–1768.

1773. (*Tölpel*)—Muller (P. L. S.) Vollständiges Natursystem. 6 vols. 8vo. Nürnberg, 1773–76; vol. 2.

*1778. (*Dronte*)—Morel ( ). Sur les Oiseaux monstrueux nommés Dronte, Dodo, Cygne capuchonné, Solitaire, et Oiseau de Nazare, et sur la petite Isle de Sable à 50 lieues environ de Madagascar, in "Observations sur la Physique." vol. 12. p. 154.

1779. (*Didus ineptus*)—Blumenbach (J. F.) Handbuch der Naturgeschichte. 8vo. Göttingen, 1779.—— (French) Manuel d'Histoire naturelle, trad. par. *S. Artaud.* 2 vols. 8vo. Metz, 1803; vol. 1. p. 256. pl. .——(English) tr. by *R. T. Gore.* 8vo. London, 1825, p. 119.

*1781. (*Dronte, Tölpel, Nazarvogel*)—Borowski (G. H.) Gemeinnützige Naturgeschichte des Thierreichs. 2 vols. 8vo. Berlin u. Stralsund. 1781, 1782; vol. 1. pp. 161, 162. pl. 25.

*1783. (*Didus ineptus*)—Hermann (J.) Tabula affinitatum Animalium. 4to. Argentorati, 1783, pp. 132, 163.

1784. (*Dronte*)—Leske (N. G.) Anfangsgründe der Naturgeschichte, ed. 2. 8vo. Leipzig, 1784.

*1785. (*Hooded Dodo* and *Nazarene Dodo.*)—Latham (John). A General Synopsis of Birds. 3 vols. 4to. London, 1781–1785; vol. 3. pp. 1, 4. pl. 70.—Sup. 2. p. 286.

*1788. (*Didus ineptus* and *D. Nazarenus*)—Gmelin (J. F.) *Caroli à Linné* Systema Naturæ, editio decima tertia, aucta, reformata. 3 vols. 8vo. Lipsiæ, 1778–1793; vol. 1. p. 728.

*1788. (*Dronte* and *Oiseau de Nazare*)—Ray (P.A.F.) Zoologie universelle et portative. 4to. Paris, 1788, pp. 188, 386.

1788. (*Dronte*)—Batsch (A. J. G. K.) Versuch einer Anleitung zur Kenntniss und Geschichte der Thiere und Mineralien, für akademische Vorlesungen entworfen. 2 vols. 8vo. Jena, 1788–89.

1789. (*Tölpel*)—Bechstein (J. M.) Gemeinnützige Naturgeschichte Deutschlands. 4 vols. 8vo. Leipzig, 1789–95.

1790. (*Dronte*)—Funke (C. P.) Naturgeschichte und Technologie. 8vo. Braunschweig, 1790.

(*Dronte, Ebelvogel, Mönchschwan*)—Gatterer (C. W. J.). Vom Nutz u. Sch. d. Th.

1790. ( )—Blumenbach (J. F.) Beyträge zur Naturgeschichte. 2 vols. 8vo. Göttingen, 1790; vol. 1. p. 24.

*1790. (*Didus ineptus* and *D. nazarenus*)—Bonnaterre (L'Abbé). Tableau encyclopédique et méthodique des trois règnes de la Nature. Ornithologie. 3 vols. 4to. Paris, 1790—1823; vol. 1. pp. 166, 167.

*1790. (*Didus ineptus*, and *D. nazarenus*)—Latham (John). Index Ornithologicus sive Systema Ornithologiæ. 2 vols. 4to. London, 1790, pp. 662, 663.

1792. (*Tölpel*)—Bechstein (J. M.) Kurzgefasste Naturgeschichte des In- u. Auslandes. 2 vols, 8vo. Leipzig, 1792–94; v. 1. p. 456.

*1792. (*Didus ineptus*)—Shaw (George). Naturalist's Miscellany, or coloured figures of Natural Objects drawn and described immediately from Nature. 24 vols. 8vo. London, 1790–1813; vol. 4. pl. 123, 143; vol. 5. pl. 166.

1793. (*Dronte*)—Donndorff (J. A.) Handbuch der Thiergeschichte. 8vo. Leipzig, 1793.

*1795. (*Didus ineptus*)—Donndorff (J. A.) Ornithologische Beyträge zur xiii. Ausgabe des Linneischen Natursystems. 2 vols. 8vo. Leipzig, 1795; vol. 2. p. 19.

*1795. (*Gemeine Dudu* and *Nazarene Dudu*)—Bechstein (J. M.) *Johann Latham's* allgemeine Uebersicht der Vögel. 4 vols. 4to. Nürnberg, 1792–1812; vol. 2. pp. 764, 766. pl. 71.

1796. ( )—Blumenbach (J. F.) Abbildungen der Naturhistorische Gegenstände. Göttingen, 1796–1810, pl. 35.

*1798. (*Dronte*)—Cuvier (George). Tableau élémentaire de l'Histoire Naturelle des Animaux. 1 vol. 8vo. Paris, An. vi., p. 251.

*1801. (*Didus ineptus*)—Stewart (C.) Elements of Natural History. 2 vols. 8vo. Edinburgh, 1801; vol. 1. p. 233.—Second edition. 2 vols. 8vo. Edinburgh, 1817; vol. 1. p. 219.

*1801. (*Dodo*)—Grant (Charles). The History of Mauritius, or the Isle of France, and the neighbouring Islands from their first discovery to the present time. 4to. London, 1801, p. 144.*

*1804. (*Dodo*)—Bory St. Vincent (J. B. G. M.) Voyage dans les quatre principales Iles des Mers d'Afrique. 3 vols. 8vo. Paris, 1804, vol. 2. p. 302.

*1806. (*Dronte*)—Dumeril (Constant). Zoologie analytique ou méthode naturelle de classification des Animaux. 8vo. Paris, 1806, p. 56.

*1808. (*Didus ineptus* and *D. nazarenus*)—Rees (Abraham). Article Didus in "The New Cyclopædia or Universal Dictionary of Arts, Sciences, and Literature." Vol. 10. pt. 2.

1809. (*Dodo*)—Shaw (George). Zoological Lectures delivered at the Royal Institution. 2 vols. 8vo. London, 1809; vol. 1. p. 213. pl. 69.

*1811. (*Didus ineptus*)—Illiger (Carolus). Prodromus Systematis Mammalium et Avium, additis terminis zoographicis utriusque classis eorumque versione germanica. 8vo. Berolini. 1811, p. 245.

*1817. (*Dronte*)—Cuvier (George). Le Règne Animal distribué d'après son organization. 4 vols. 8vo. Paris, 1817; vol. 1. p. 463.—Nouvelle édition. 5 vols. 8vo. Paris, 1829; vol. 1. p. 497.

*1817. (*Dronte* and *Oiseau de Nazare*)—Sonnini (C. W. S.) Nouveau Dictionnaire d'Histoire Naturelle. Nouvelle édition. 36 vols. 8vo. Paris, 1816—1819; v. 9. p. 589; v. 23. p. 431.

*1819. (*Dronte*)—Dumont (C.) Article Dronte in "Dictionnaire des Sciences naturelles." 8vo. Paris; vol. 13. p. 519.

*1820. (*Didus ineptus*)—Temminck (C. J.) Manuel d'Ornithologie, ou Tableau systématique des Oiseaux qui se trouvent en Europe; précédé d'une Analyse du Système général d'Ornithologie. 2nde édition, 4 vols. 8vo. Paris, 1820–1840; pt. 1. p. cxiv.

*1823. (*Hooded Dodo* and *Nazarene Dodo*)—Latham (John). A general history of Birds. 10 vols. 4to. Winchester, 1821–1824; vol. 8. pp. 372, 375. pl. 135.

*1823. (*Didus*)—Vigors (N. A.) Observations on the Natural Affinities that connect the Orders and Families of Birds; in the "Transactions of the Linnean Society of London." vol. 14. p. 484.

*1825. (*Oiseau de Nazare*)—DUMONT (C.) Article OISEAU DE NAŻARE in "Dictionnaire des Sciences naturelles." vol. 35. p. 494.

*1826. (*Didus ineptus*)—STEPHENS (J. F.) General Zoology, or systematic Natural History, by *George Shaw*, continued by *J. F. Stephens*. 14 vols. 8vo. London, 1800–1826; vol. 14. p. 308. pl. 40.

*1827. (*Didus ineptus*)—GRAY (John Edward). On the Dodo; in the "Zoological Journal." v. 3. p. 605.

*1828. (*Didus ineptus*)—DUNCAN (John Shute). A summary review of the authorities on which naturalists are justified in believing that the Dodo, *Didus ineptus*, Linn., was a bird existing in the Isle of France, or neighbouring islands, until a recent period; in the "Zoological Journal." v. 3. p. 554.

*1828. (*Didus ineptus*)—ESTRUP (P. J.) Haandbog i Ornithologien eller Naturhistorie of de mærkværdigste Fugle. 8vo. Kiöbenhavn, 1828; p. 173.

*1828. (*Didus ineptus*)—STARK (John). Elements of Natural History. 2 vols. 8vo. Edinburgh, 1828; vol. 1. p. 330.

*1828. (*Dronte*)—LESSON (R. P.) Manuel d'Ornithologie ou description des genres et des principales espèces d'oiseaux. 2 vols. 12mo. Paris, 1828; vol 2. p. 210.

*1829. (*Dodo*)—GRIFFITH (Edward). The Animal Kingdom arranged in conformity with its organization, by *Baron Cuvier*; with additional descriptions by *E. Griffith*. 16 vols. 8vo. London, 1827–1835; vol. 8. pp. 299, 443.

*1829. (*Dodo*)—THOMPSON (J. V.) Contributions towards the Natural History of the Dodo (*Didus ineptus*), a bird which appears to have become extinct towards the end of the 17th or beginning of the 18th century; in *Loudon's* "Magazine of Natural History." vol. 2. p. 443.

*1830. (*Dodo*)—BLAINVILLE (H. D. de). Mémoire sur le Dodo, autrement Dronte, in "Nouvelles Annales du Muséum d'Histoire Naturelle." vol. 4. p. 1. pl. 1–4.

*1831. (*Didus ineptus*)—EICHWALD (Edward). Zoologia specialis. 3 vols. 8vo. Vilnæ, 1831; vol. 3. p. 257.

*1832. (*Didus*)—BOIE (F.) Art. DIDUS in Ersch and Gruber's "Allgemeine Encyclopädie der Wissenschaften u. Kunste." 4to. Leipzig; vol. 24. p. 545.

*1833. (*Dodo*)—LYELL (Charles). Principles of Geology. 2nd ed. 3 vols. 8vo. London, 1833; v. 2. p. 157. —3rd ed. 4 vols. 12mo. London, 1834; v. 3. p. 60.

*1833. (*Dodo*)—KNIGHT (Charles). On the Dodo, in the "Penny Magazine." 8vo. London, 1832–1846; vol. 2. p. 209.

*1835. (*Dodo*)—SWAINSON (W.) Treatise on the Geography and Classification of Animals, in *Lardner's* "Cabinet Cyclopædia." p. 112.

*1836. (*Dodo*)—WIEGMANN (A. F. A.) Ueber den Dodo; in "*Weigmann's* Archiv für Naturgeschichte.' 1836; v. 2. p. 271.

*1836. (*Dronte*)—KAUP (J. J.) Das Thierreich in seinen Hauptformen systematisch beschrieben. 3 vols. 8vo. Darmstadt, 1835, 1836; vol. 2. p. 232.

*1836. (*Dodo*)—BUCKLAND (William). Geology and Mineralogy considered with reference to Natural Theology. 2 vols. 8vo. London, 1836; vol. 2. p. 17. pl. 1. f. 120.

*1837. (*Didus*)—BRONN (H. G.) Lethæa Geognostica, oder Abbildungen u. Beschreibungen der für die Gebirgsformationen bezeichnendsten Versteinerungen. 2 vols. 8vo. and atlas 4to. Stuttgart, 1835–1837; pp. 824, 1171. pl. 44. f. 7.

*1837. (*Dodo*)—BRODERIP (William John). The article DODO in the "Penny Cyclopædia." vol. 9. p. 47.

*1839. (*Didus ineptus*)—LA FRESNAYE (M. de). Nouvelle Classification des Oiseaux de Proie, ou Rapaces; in the "Revue Zoologique par la Société Cuvierienne." 1839. p. 193.

*1840. (*Dodo*)—GRAY (J. E.) Synopsis of the contents of the British Museum. 12mo. London, 1840. p. 99.

1841. (*Dronte*)—REINHARDT (Cand.) in Froriep's "Notizen." 1841. No. 364.

*1842. (*Dronte*)—REINHARDT (Cand.) Nöiere Oplysning om det i Kiöbenhavn fundne Drontehoved; in *H. Kröyer's* "Naturhistorisk Tidskrift." 8vo. Kiöbenhavn; vol. 4. p. 71.

*1842. (*Dodo*)—Owen (Richard) Notice of Savery's picture at the Hague in the "Penny Cyclopædia." vol. 23. p. 143.

*1843. (*Didus ineptus*)—Lehmann ( ). Ein Nachtrag über den Didus ineptus. 8vo. Kopenhagen; 1843.—Nov. Act. Ac. Leop. Car.; vol. 21. p. 1.

*1844. (*Dodo*)—Strickland (H. E.) Report on the recent progress and present state of Ornithology in "Reports of the British Association for the Advancement of Science" for 1844, p. 213.

*1845. (*Didus ineptus*)—Owen (R.) Descriptive and illustrated Catalogue of the fossil organic remains of Mammalia and Aves contained in the Museum of the Royal College of Surgeons. 4to. London, 1845, p. 339.

*1845. (*Didus*)—Hamel (J.) Ueber *Dinornis* und *Didus*, zwei ausgestorbene Vogelgattungen; in "Bulletin de la Classe physico-mathématique de l'Académie Impériale des Sciences de Saint-Pétersbourg." vol. 4. p. 49.

*1846. (*Didus ineptus*)—Owen (R.) Observations on the Dodo, in "Transactions of the Zoological Society of London." vol. 3. pp. 331, 335.—Proceedings of the Zoological Society, part 14. p. 51.

*1846. (*Didus ineptus*)—Carus (C. G.) England u. Schottland im Jahre 1844, vol. 1. p. 375.—(English) The King of Saxony's Journey through England and Scotland in the year 1844. (Trans. by *S. C. Davison.*) 8vo. London, 1846; p. 187.

*1846. (*Dodo*)—Hamel (J.) Sur un Crâne de Dodo au Musée de Copenhague; in "Bull. Classe phys. math. Ac. Imp. Sc. St. Pétersb." vol. 5. p. 314.—Instit. no. 709. p. 252.—Edinb. New Phil. Journ. v. 43. p. 405.

*1847. (*Dodo*)—Hamel (J.) Tradescant der Aeltere, 1618 in Russland. 4to. Petersburg. 1847; p. 169.—Recueil des Actes de la Séance publique de l'Académie Impériale de St. Pétersbourg.

*1847. (*Dodo*)—Strickland (H. E.) On the history of the Dodo and other allied species of Birds, in "Reports of the British Association" for 1847, Sections, p. 79.—Athenæum, 1847, pp. 747, 769.

*1847. (*Do-do*)—Forbes (Edward). The fate of the Dodo, an ornithological Romance; in the "Literary Gazette," July 3, 1847, p. 493.

*1847. (*Didus ineptus*)—Sundevall (C. J.) Arsberättelse om Zoologiens Framsteg under ären 1843 och 1844. 8vo. Stockholm 1847; p. 183.

1847. (*Dodo*)—Canino (C. L. Bonaparte, Prince of). On the Dodo.—Riunione degli Scienziati Italiani in Venezia, Settembre, 1847.—Allgemeine Zeitung, Sept. 24, 1847; Beilage, p. 2131.

*1848. (*Dodo*)—Brandt (J. F.) Untersuchungen über die Verwandschaften, die systematische Stellung, die geographische Verbreitung und die Vertilgung des Dodo, nebst Bemerkungen über die im Vaterlande des Dodo oder auf den Nachbarinseln desselben früher vorhandenen grossen Wadvögel.—Bulletin de la Classe phys. math. de l'Acad. Imp. de St. Pétersbourg. vol. 7. p. 38.

*1848. (*Dronte*)—Fitzinger (L. J.) Mittheilungen über eine Original-Abbildung des Dronte, (*Didus ineptus*, Linné) von *Roland Savery* in der k. k. Gemälde-Gallerie im Belvedere zu Wien; in *Wiegmann's* "Archiv für Naturgeschichte," 1848. p. 79.

*1848. (*Dodo*)—Hamel (J.) Der Dodo, die Einsiedler, und die erdichtete Nazarvogel. 8vo. Petersburg, 1848.—Bulletin physico-mathématique de l'Académie des Sciences de St. Pétersbourg. vol. 7. No. 5, 6.

*1848. (*Dodo*)—Strickland (H. E.) and Melville (A. G.) The Dodo and its Kindred, or the History, Affinities, and Osteology of the Dodo, Solitaire, and other extinct Birds of the Islands Mauritius, Rodriguez, and Bourbon. 4to. London, 1848.

II. THE SOLITAIRE OF RODRIGUEZ.

*1691. (*Solitaire*)—Leguat (François). Voyages et Avantures de *François Leguat*. 2 vols. 12mo. London, 1708; ed. 2. 1720.—(English) A new Voyage to the East Indies, by *Francis Leguat* and his companions. 12mo. London, 1708.

*1770. (*Solitaire*)—BUFFON (G. L. Le Clerc de). Histoire naturelle des Oiseaux. 9 vols. 4to. Paris, 1770–1783; vol. 1. p. 485.—Ed. 2. 10 vols. fol. Paris, 1771–1786; vol. 2. p. 77.—Nouvelle édition par *C. S. Sonnini*, 28 vols. 8vo. Paris, 1801–1805; vol. 4. p. 343. pl. 33. f. 2.

*1781. (*Einsiedler*)—BOROWSKI (G. H.) Gemeinnützige Naturgeschichte des Thierreichs. 2 vols. 8vo. Berlin 1781, 1782; vol. 1. p. 162.

*1785. (*Solitary Dodo*)—LATHAM (John). A general Synopsis of Birds. 3 vols. 4to. London, 1781–1785; vol. 3. p. 3.

*1788. (*Didus solitarius*)—GMELIN (J. F.) *Caroli à Linné* Systema Naturæ, editio decima tertia, aucta, reformata. 3 vols. 8vo. Lipsiæ, 1788–1793; vol. 1. p. 728.

*1788. (*Solitaire*)—RAY (P. A. F.) Zoologie universelle et portative. 4to. Paris, 1788, p. 567.

*1790. (*Didus solitarius*)—BONNATERRE (L'Abbé). Tableau encyclopédique et méthodique des trois règnes de la Nature. Ornithologie. 3 vols. 4to. Paris, 1790–1823; vol. 1. p. 166.

*1790. (*Didus solitarius*)—LATHAM (John). Index Ornithologicus sive Systema Ornithologiæ. 2 vols. 4to. London, 1790, p. 662.

*1795. (*Didus solitarius*)—DONNDORF (J. A.) Ornithologische Beyträge zur xiii. Ausgabe des Linneischen Natursystems. 2 vols. 8vo. Leipzig, 1795; vol. 2. p. 20.

*1795. (*Didus solitarius*)—BECHSTEIN (J. M.). *Johann Latham's* allgemeine Uebersicht der Vögel. 4 vols. 4to. Nürnberg, 1792–1812; vol. 2. p. 765.

*1801. (*Solitaire*)—GRANT (Charles). The History of Mauritius, or the Isle of France, and the neighbouring Islands, from their first discovery to the present time. 4to. London, 1801; p. 117.

*1808. (*Didus solitarius*)—REES (Abraham). Article DIDUS in "The New Cyclopædia or Universal Dictionary of Arts, Sciences, and Literature." Vol. 10. pt. 2.

*1819. (*Solitaire*)—SONNINI (C. N. S.) Nouveau Dictionnaire d'Histoire Naturelle. Nouvelle edition. 36 vols. 8vo. Paris, 1816–1819; vol. 31. p. 376.

*1823. (*Solitary Dodo*)—LATHAM (John). A general History of Birds. 10 vols. 4to. Winchester, 1821–1824; vol. 8. p. 374.

*1827. (*Solitaire*)—DUMONT (C.) Art. SOLITAIRE in "Dictionnaire des Sciences naturelles." Vol. 49. p. 451.

*1830. (*Dronte*)—CUVIER (G.) Sur quelques ossemens qui paraissent appartenir à une espèce perdue seulement depuis deux siècles. Ann. des Sc. Nat. vol. 21. Rev. Bibl. p. 103.—Revue Sept. 103, 104, 109, 110. —Bull. Sc. Nat. vol. 22. p. 122.—Edinb. Journ. Nat. Sc. vol. 3. p. 30.

*1832. (*Dodo*)—DESJARDINS (Julien). Analyse des travaux de la Société d'Histoire Naturelle de l'Ile Maurice pendant la 2de Année.—Proc. Com. Zool. Soc. pt. 2. p. 111.—Phil. Mag. ser. 2. v. 1. p. 461.

*1833. (*Dodo*)—TELFAIR (Charles). On bones of the Dodo found in Rodriguez, in "Proceedings of Zoological Society of London," part 1. p. 31.

*1844. (*Solitaire*)—STRICKLAND (H. E.) On the evidence of the former existence of Struthious Birds, distinct from the Dodo, in the islands near Mauritius; in "Proceedings of the Zoological Society of London," part 12. p. 77.

III. BREVIPENNATE BIRDS IN BOURBON.

*1613. (*A great fowl*)—TATTON (J.) Voyage of *Castleton* in *Purchas's*, Pilgrimage. ed. 1625, vol. 1. p. 333.—*Prévost*, Histoire générale des Voyages, vol. 2. p. 120.—*Harris's* Voyages, vol. 1. p. 115.—*Grant's* Mauritius, p. 164.

*1618. (*Dod-eersen*)—BONTEKOE VAN HOORN (W. Y.) Journael ofte gedenckwaerdige beschrijvinge van de Oost-Indische Reyse. 4to. Haerlem, 1646, p. 6; Rotterdam, 1647, p. 7; Amsterdam, 1648, p. 5; 1650, p. 5; 1656; Utrecht, 1649, p. 6; 1651.—Journael van de acht-jarige avontuerlijcke Reyse van *Willem*

Ysbrantsz Bontekoe van Hoorn, gedaen nae Oost-Indien. 4to. Amsterdam, by *Gillis Joosten Zaagman.* No date.—(French) in *Thevenot's* Relations de divers Voyages curieux. Paris, 1663, vol. 1.—(German) in *Hulsius*, Vier und zwanzigste Schiffart. 4to. Francfort, 1648. p. 7.

*1668. (*Oiseau Solitaire*)—CARRE (M.) Voyages des Indes Orientales. 2 vols. 12mo. Paris, 1699; vol. 1. p. 12. —*Prévost*, Hist. gén. des Voyages, vol. 9. p. 3.

*1669. (*Solitaire et Oiseau bleu*)—D. B. (Sieur). MS. Journal in Library of Zoological Society.—Proceedings of Zool. Soc. pt. 12. p. 77.

*1819. (*Oiseau bleu*)—REES (A.) Cyclopædia, art. "*Bourbon.*"

*1829. (*Dronte ou Solitaire*)—BILLIARD (A.) Voyage aux Colonies Orientales, ou lettres écrites des Isles de France et de Bourbon, pendant les années 1817, 1818, 1819, et 1820. 8vo. Paris, 1829, p. 261.

*1844. (*Solitaire* and *Oiseau bleu*)—STRICKLAND (H. E.) On the evidence of the former existence of Struthious Birds distinct from the Dodo in the islands near Mauritius; in "Proceedings of the Zoological Society of London," part 12. p. 78.

LIST OF PLATES.

Plates II., III., III.*, IV., and IV.*, are examples of various applications of *Anastatic Printing*. Plate II. is a fac-simile of an engraving executed by tracing the original, line for line, with a steel pen, lithographic ink, and tracing paper. The drawing is then transferred, by the Anastatic process, to a plate of zinc, and printed from as in ordinary zincography or lithography. Plate IV. is executed in the same way as Plate II., except that its details are copied by the eye instead of being *traced*. Plates III., III.* and IV.*, are examples of a new art to which I have given the name of *Papyrography*, (See *Athenæum*, Feb. 12, 1848.) It consists in drawing on paper with *lithographic chalk*, and in transferring the drawings, so made, to a plate of zinc, by the Anastatic process. These drawings, when printed, bear a close resemblance to lithographs, and enable an artist or a traveller by merely using *lithographic chalk* instead of a *lead pencil*, to print and publish his original sketches (without *redrawing* or *reversing*), at any interval of time. For Plate III.* and IV.* I am indebted to E. Higgin, Esq., of Liverpool, who sent the drawings by post to Oxford, where they were transferred and printed by Mr. P. H. Delamotte.—H. E. S.

PLATE V.

Fig. 1. Side view of the head of the DODO, with the dried skin, from the unique specimen in the Ashmolean Museum at Oxford.

Fig. 2. Side view of the head of the DODO, restored chiefly from the celebrated picture, presented by Edwards to the British Museum. The great development of the cere, the tubular nostril opening forwards, the form and abrupt termination of the horny sheaths which have disappeared in Fig. 1, the extent of the gape, and the caruncular folds at the base of the upper gnathotheca, on the forehead, and extending from the angle of the mouth, are well exhibited.

PLATE VI.

Front, side, and back views of the leg of the DODO, in the British Museum. These two plates were executed for that valuable work, the "Genera of Birds," by Messrs. G. R. Gray, and D. W. Mitchell, who have obligingly allowed us the use of them.

PLATE VII.

Fig. 1. *Didunculus strigirostris*, one-third of natural size, (reduced from the figure in Mr. Gould's "Birds of Australia.")

Fig. 2. Head of ditto, natural size, to show the extension of the cere round the eye, the nostril opening downwards, the great curvature of the upper, and the teeth on the lower horny sheath, with the abrupt termination of both.

Fig. 3. Head of *Treron abyssinica*. The peculiar columbine cere and pouting of the nasal scale, the oblique orifice of the nostril inclined forwards and upwards, and the abrupt termination of the horny sheaths are shown.

Fig. 4. Head of *Verrulia carunculata*, to show the great development of wattles in a member of the *Columbidæ*.

Fig. 5. Front view of left leg of *Treron abyssinica*.

Fig. 5 *a*. Side view of ditto.

Fig. 6. Front view of ditto of *Geophaps scripta*.

Fig. 6 *a*. Side view of ditto.

In the former the inner toe is shorter, in the latter longer, than the outer.

PLATE VIII.

Side view of the skull of the DODO.

PLATE IX.

Fig. 1. Upper view of skull of the DODO.

Fig. 2. Lower view of ditto.

PLATE IX.*

Fig. 1. Back view of skull of the DODO.

Fig. 2. Upper view of lower jaw.

Fig. 3. Lower view of ditto.

Fig. 4. Inner view of ditto, partly in outline, as it could not be viewed directly by the artist.

Fig. 5. Circle of sclerotic bones in the DODO, with the sclerotic coat of the eye-ball.

PLATE X.

Fig. 1. Side view of skull of *Didunculus strigirostris*.

Fig. 1 *a*. Back view of ditto.

Fig. 1 *b*. Upper view of ditto.

Fig. 1 *c*. Lower view of ditto.

Fig. 2. Side view of skull of the DODO, reduced to one-third for more accurate comparison.

Fig. 2 *a*. Back view of ditto, similarly reduced.

Fig. 3. Side, Fig. 3 *a*. back, 3 *b*. upper, and 3 *c*. lower views of skull of *Treron chlorigaster*.

Fig. 4. ditto. 4 *a*. — 4 *b*. — and 4 *c*. ———————— *Goura Steursii*.

Fig. 5. ditto. 5 *a*. — 5 *b*. — and 5 *c*. ———————— *Geophaps Smithii*.

Fig. 6. Section of skull of *Treron chlorigaster*, to show the great development of the frontal diploë, which in the DODO forms the inter-orbital protuberance.

Fig. 7. Outer view of tympanic bone of *Didunculus strigirostris*.

Fig. 7 *a*. Inner view of ditto.

Fig. 7 *b*. Lower view of ditto.

Fig. 8. Articular surface of lower jaw, of *Didunculus strigirostris*.

Fig. 9. Front view of metatarsus of *Didunculus strigirostris*.

Fig. 9 *a*. Back view of ditto.

Fig. 9 *b*. Inner view of metatarsus of *Didunculus strigirostris*.
Fig. 9 *c*. Outer view of ditto.
Fig. 9 *d*. View of upper extremity of ditto.
Fig. 9 *e*. ——— lower extremity of ditto.
Fig. 10. Back view of posterior metatarsal of ditto.
Fig. 10 *a*. Front view of ditto.

PLATE XI.

Fig. 1. Front view of metatarsus of the DODO.
Fig. 2. Back view of ditto.
Fig. 3. Inner view of the metatarsus, and of posterior metatarsal of ditto.
Fig. 4. Outer view of the metatarsus of the DODO.
Fig. 5. View of upper extremity of ditto.
Fig. 6. View of lower extremity of ditto.
Fig. 7. Back view of posterior metatarsal of the DODO.
Fig. 8. Front view of ditto.
Fig. 9. Oblique view of ditto.
Fig. 10. Upper view of ditto.
Fig. 11. Front view of metatarsus of *Goura coronata*.
Fig. 12. Back view of ditto.
Fig. 13. Inner view of ditto.
Fig. 14. View of upper extremity of ditto.
Fig. 15. View of lower extremity of ditto.
Fig. 16. Back view of posterior metatarsal of ditto.
Fig. 17. Front view of ditto.
Fig. 18 and 19. Front and back views of posterior metatarsal of *Didunculus strigirostris*.
Fig. 20 to 24. Corresponding views of metatarsus of *Phaps picata*.
Fig. 25. Back view of posterior metatarsal of ditto.
Fig. 26 to 31. Corresponding views of metatarsi of *Geophaps scripta*.
Fig. 32 to 37. ——————————— of *Treron chlorigaster*.
Fig. 38 to 43. ——————————— of *Lopholæmus antarcticus*.

PLATE XII.

Fig. 1. Front view of metatarsus and toes of the DODO.
Fig. 1 *a*. Back view of ditto.
Fig. 1 *b*. Inner view of ditto.
Fig. 2. Posterior articular facets of proximal phalanges; *a*, *b*, *c*, glenoid fibro-cartilages of three front toes.
Fig. 3. Back view of outer toe of the DODO, to show the glenoid ligaments; *a*, the upper.
Fig. 4. Proximal, and 4 *a*, distal articular facets of second phalanx of outer toe of the DODO.
Fig. 5. Side view of ungual phalanx of outer toe of the DODO.
Fig. 5 *a*. Articular surface of ditto.
Fig. 6. Front view of foot of *Treron chlorigaster*.
Fig. 6 *a*. Hind toe of ditto.
Fig. 7. Front view of foot of *Columba œnas*.
Fig. 7 *a*. Hind toe of ditto.
Fig. 8. Front view of foot of *Geophaps Smithii*.
Fig. 8 *a*. Hind toe of ditto.

PLATE XIII.

PLATE XIV.

PLATE XV.

*** All the bones are figured of the natural size.

WOOD ENGRAVINGS.

INDEX.

2 o

ERRATA.

Page 4, folio, *for* iv. *read* 4.
„ 30, line 32, *for* Bellvedere, *read* Belvedere.
„ 38, „ 26, *for* 1845, *read* 1846.
„ 44, „ 9, *for* tarso-metarsal, *read* tarso-metatarsal.
„ 57, „ 27, *for* 1674, *read* 1647.
„ 70, „ 6, *for* posterior, *read* superior.
„ 70, 74, 81, *for* lacrymal, *read* lachrymal.

Plate V

Jos. Dinkel del. et lith.

Printed by Hullmandel & Walton

Plate VI.

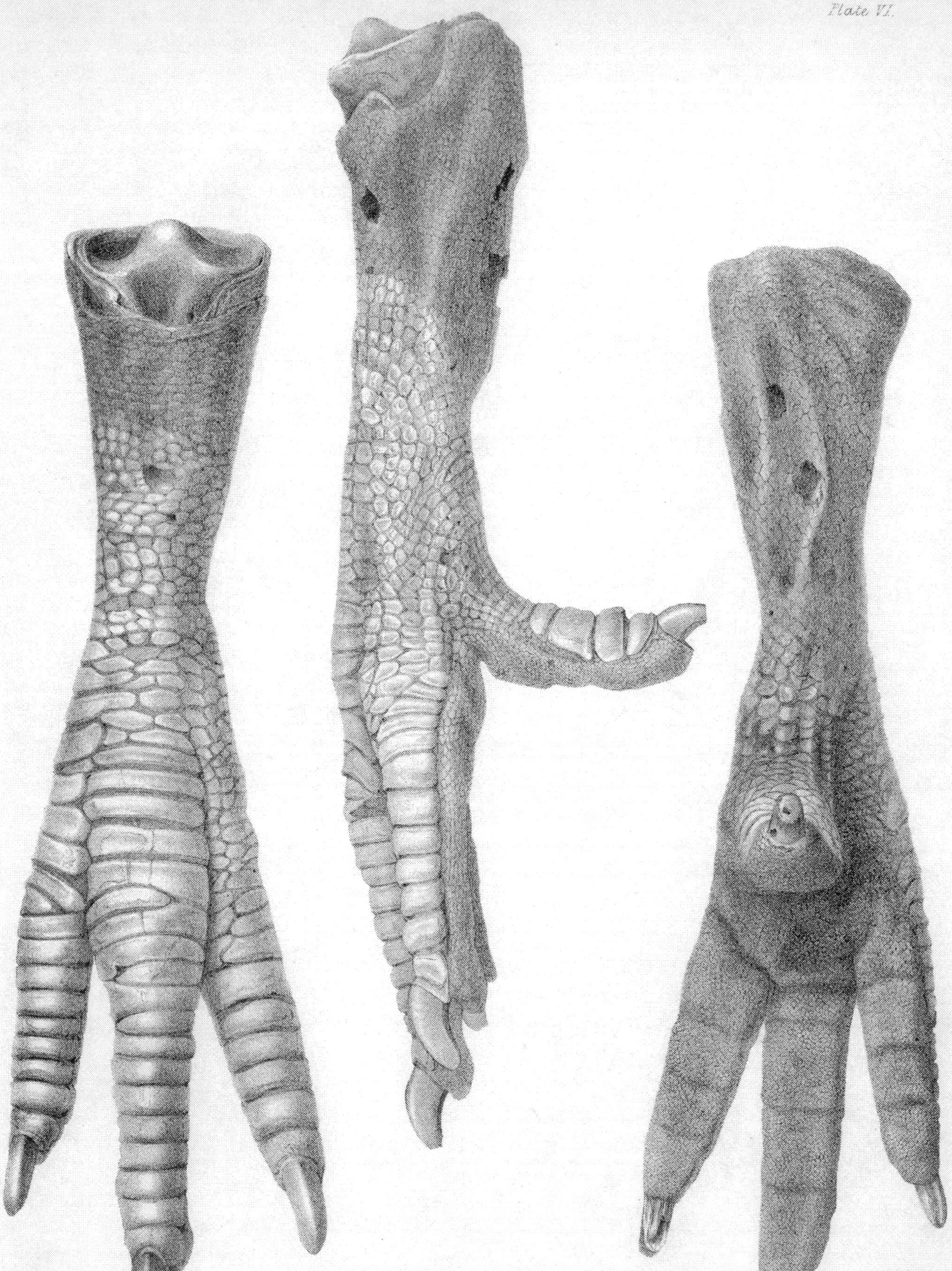

Jos. Dinkel del et lith.

Printed by Hullmandel & Walton.

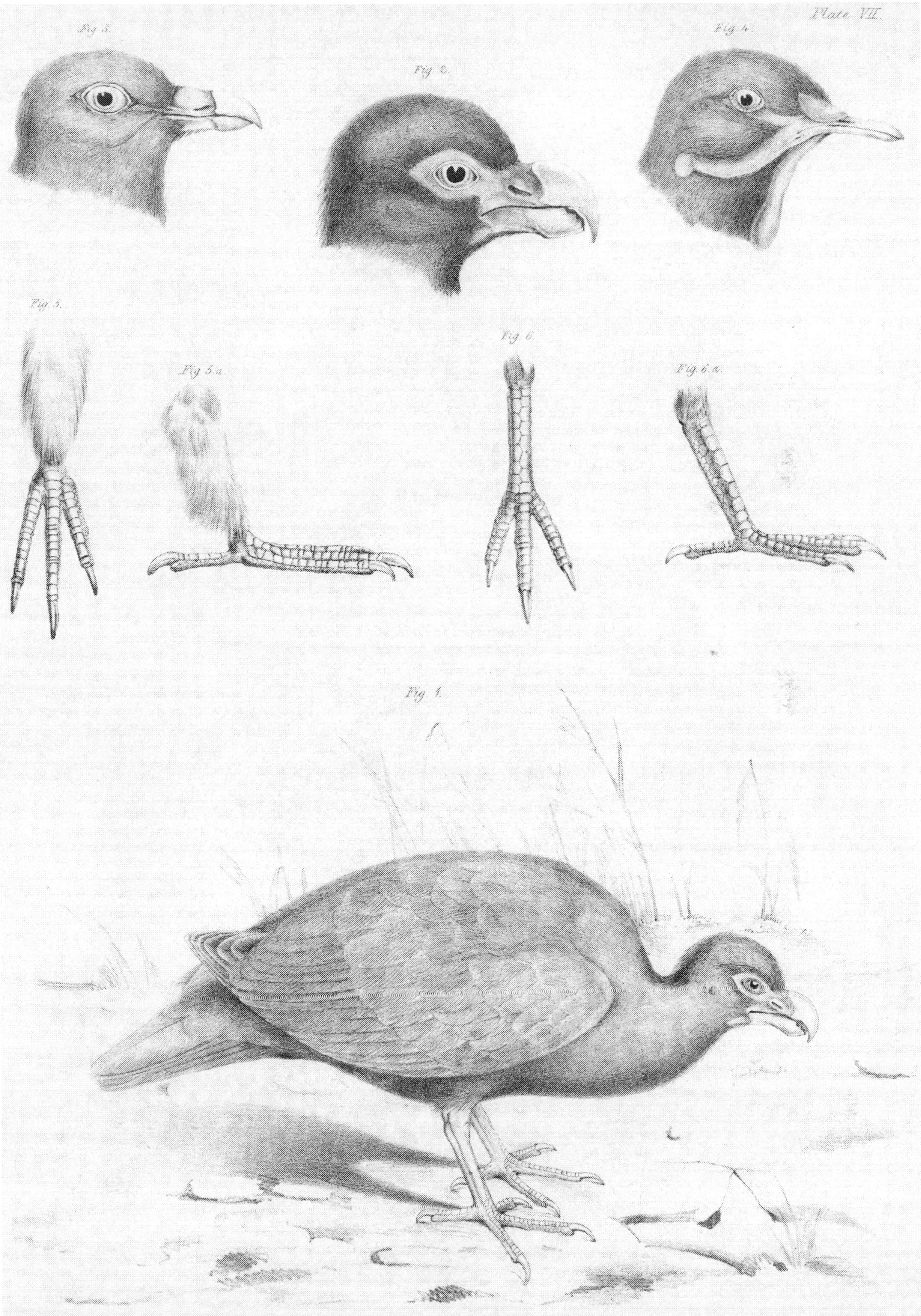
Plate VII.
Fig. 3.
Fig. 2.
Fig. 4.
Fig. 5.
Fig. 5. a.
Fig. 6.
Fig. 6. a.
Fig. 1.
Tuffen West, lith.
Reeve, Benham & Reeve, imp.

Plate VIII

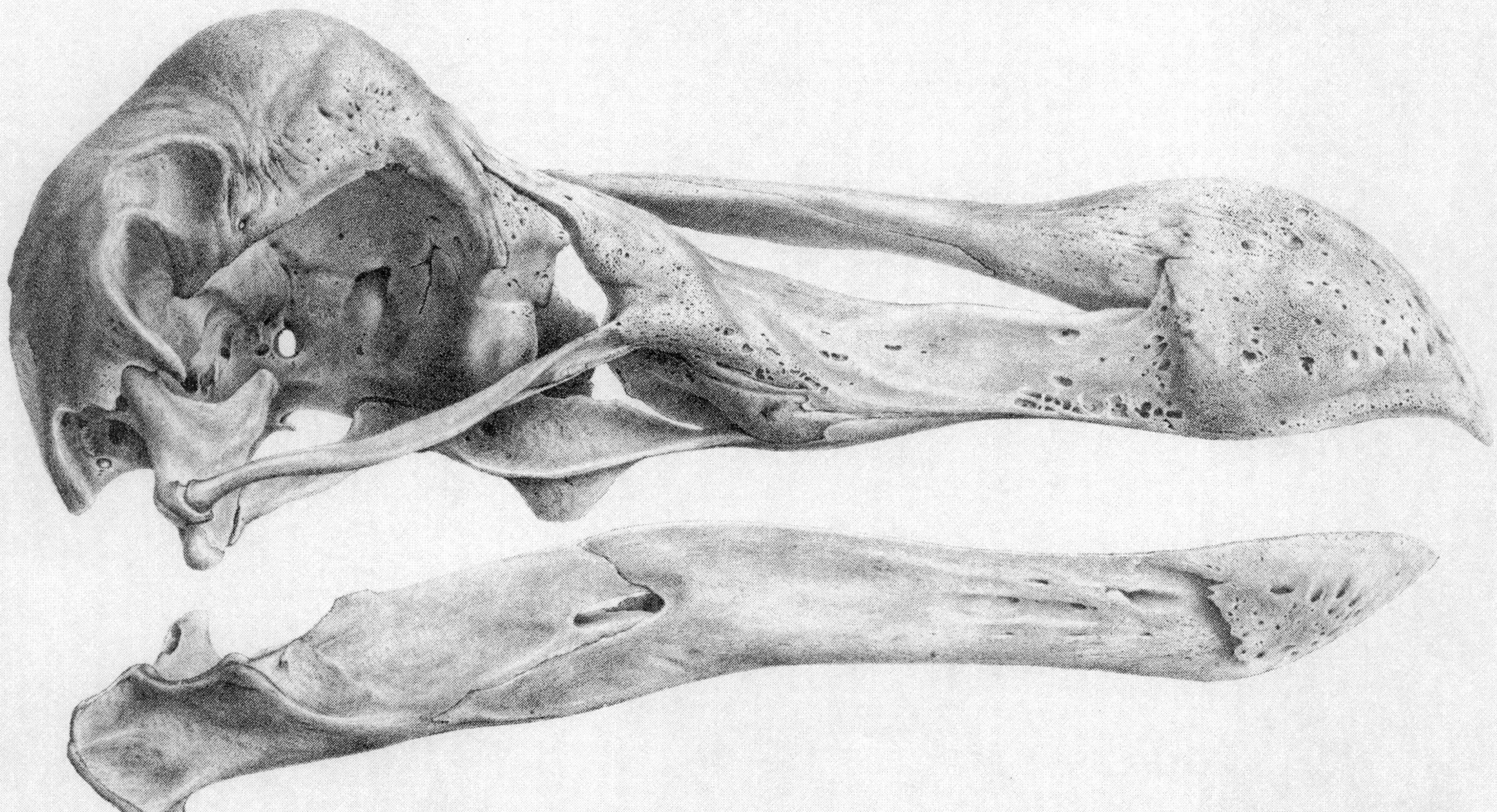

Lithographers

Plate IX.

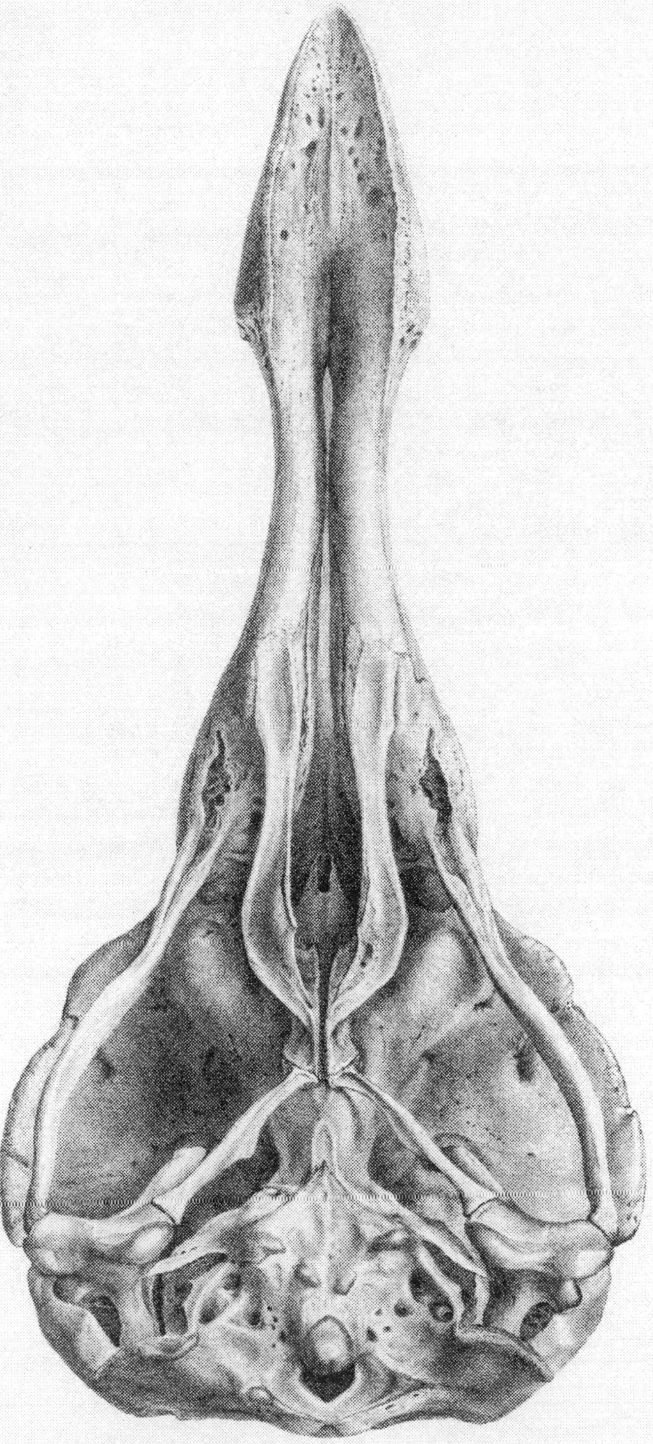

Fig. 2.

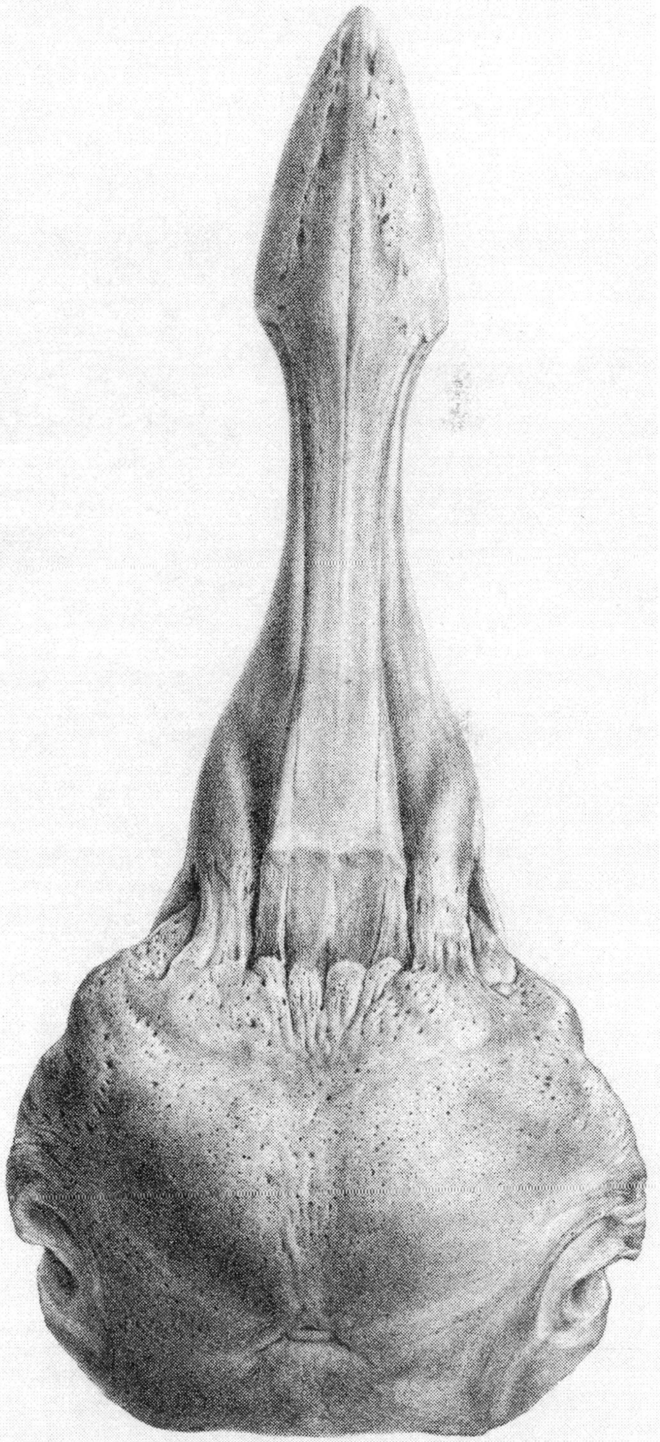

Fig. 1.

Plate IX*

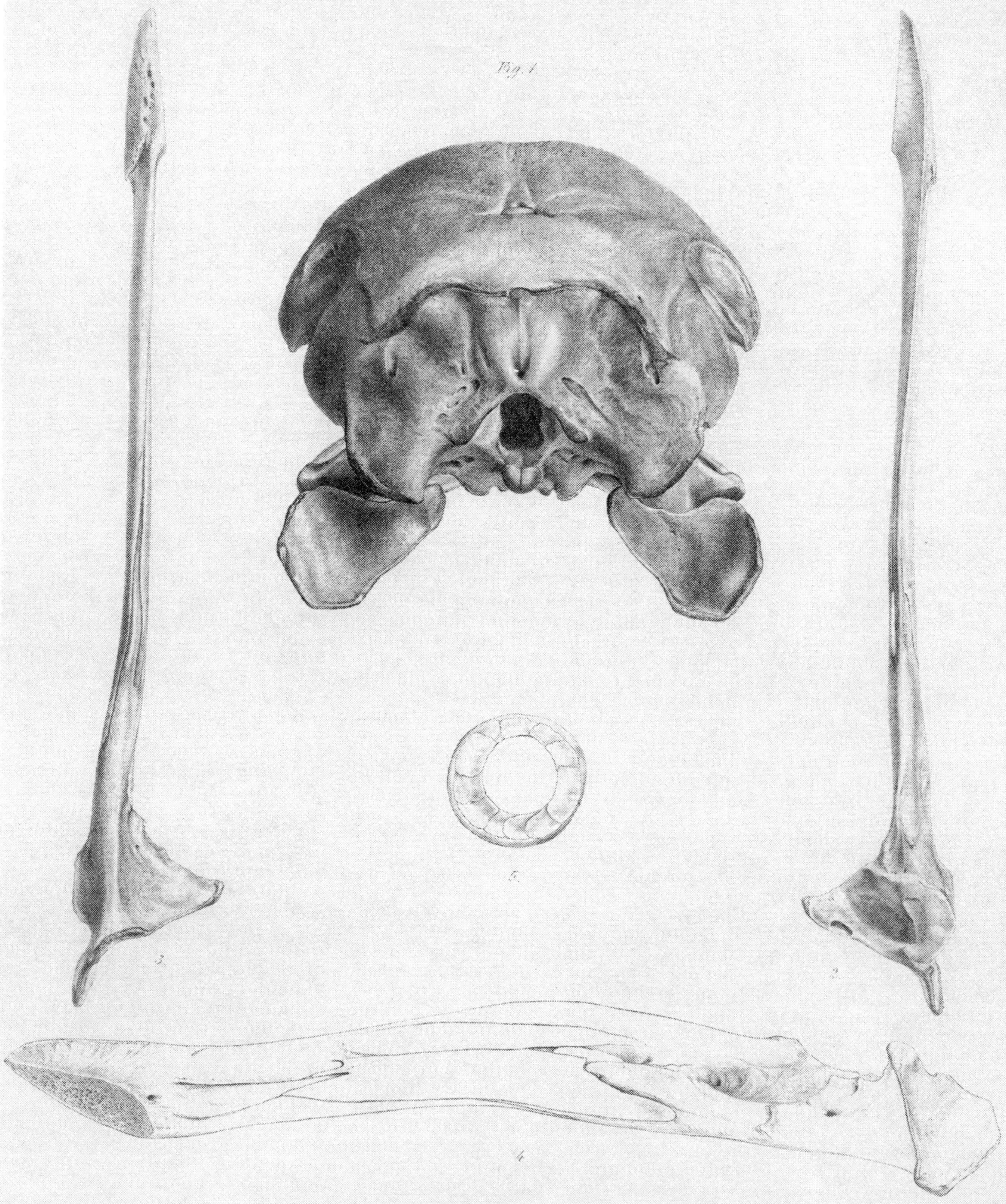

Plate X.

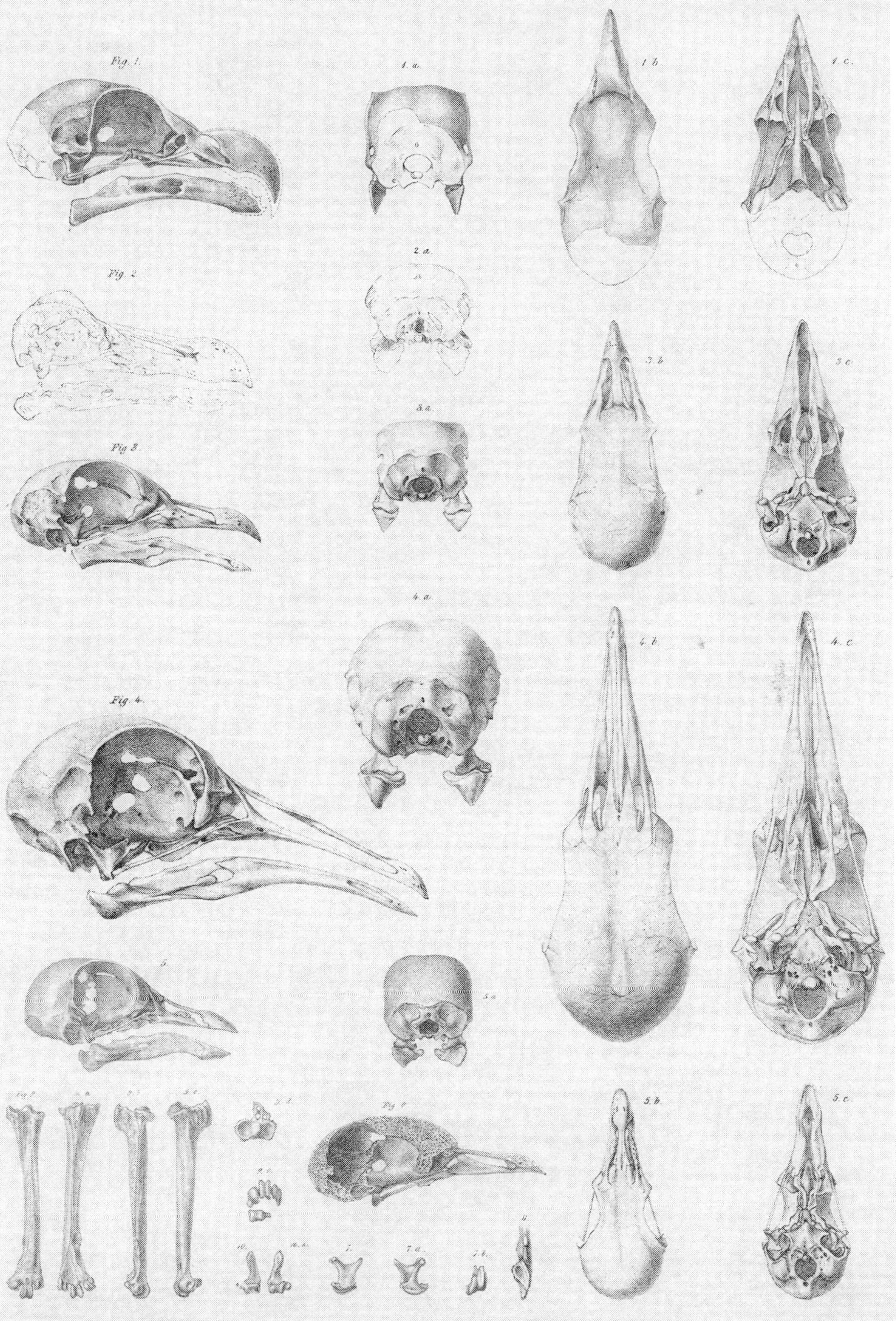

Tuffen West, lith.

Plate XI.

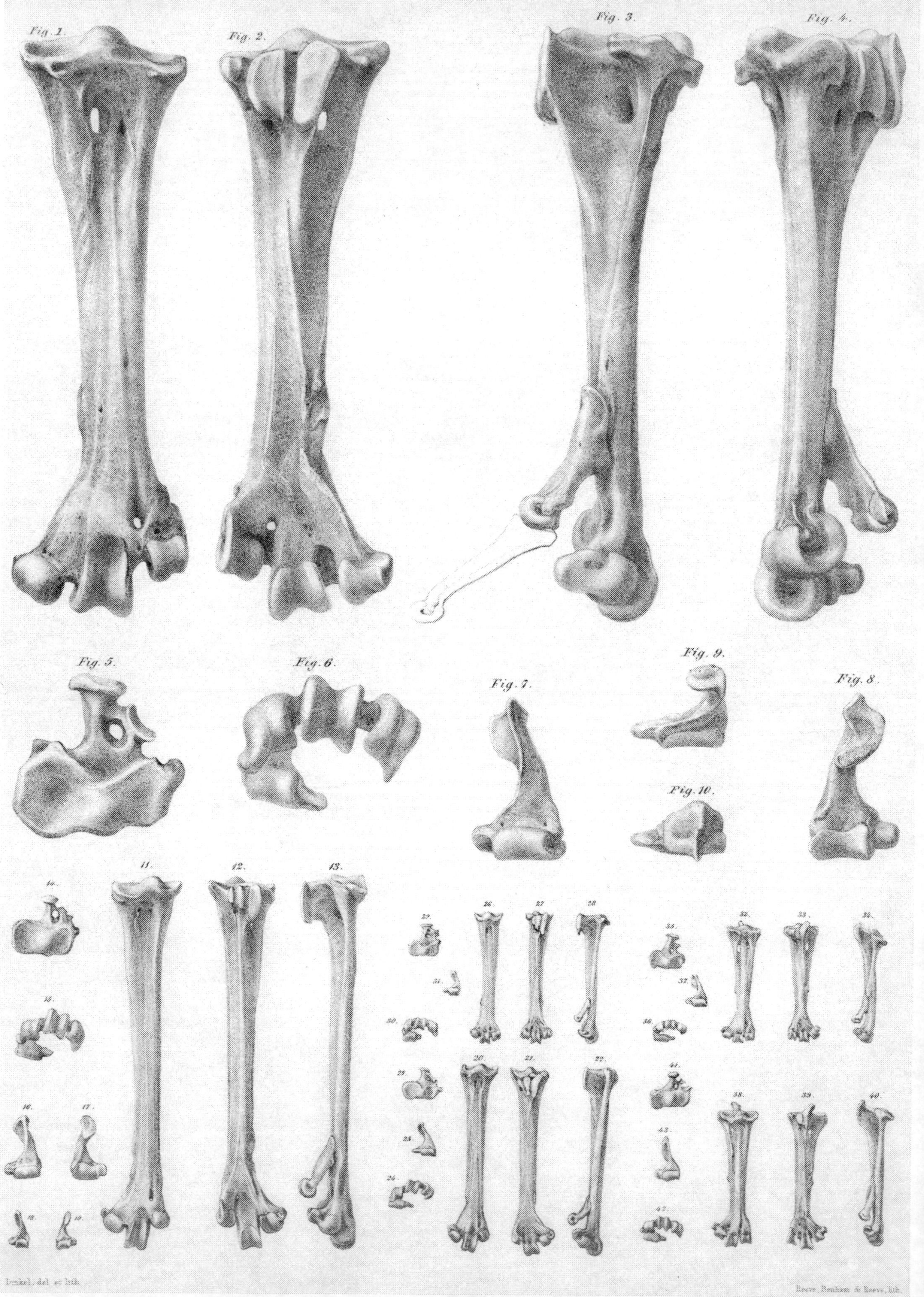

Fig. 1.
Fig. 2.
Fig. 3.
Fig. 4.
Fig. 5.
Fig. 6.
Fig. 7.
Fig. 8.
Fig. 9.
Fig. 10.
Reeve, Benham & Reeve, lith.

Plate XII.

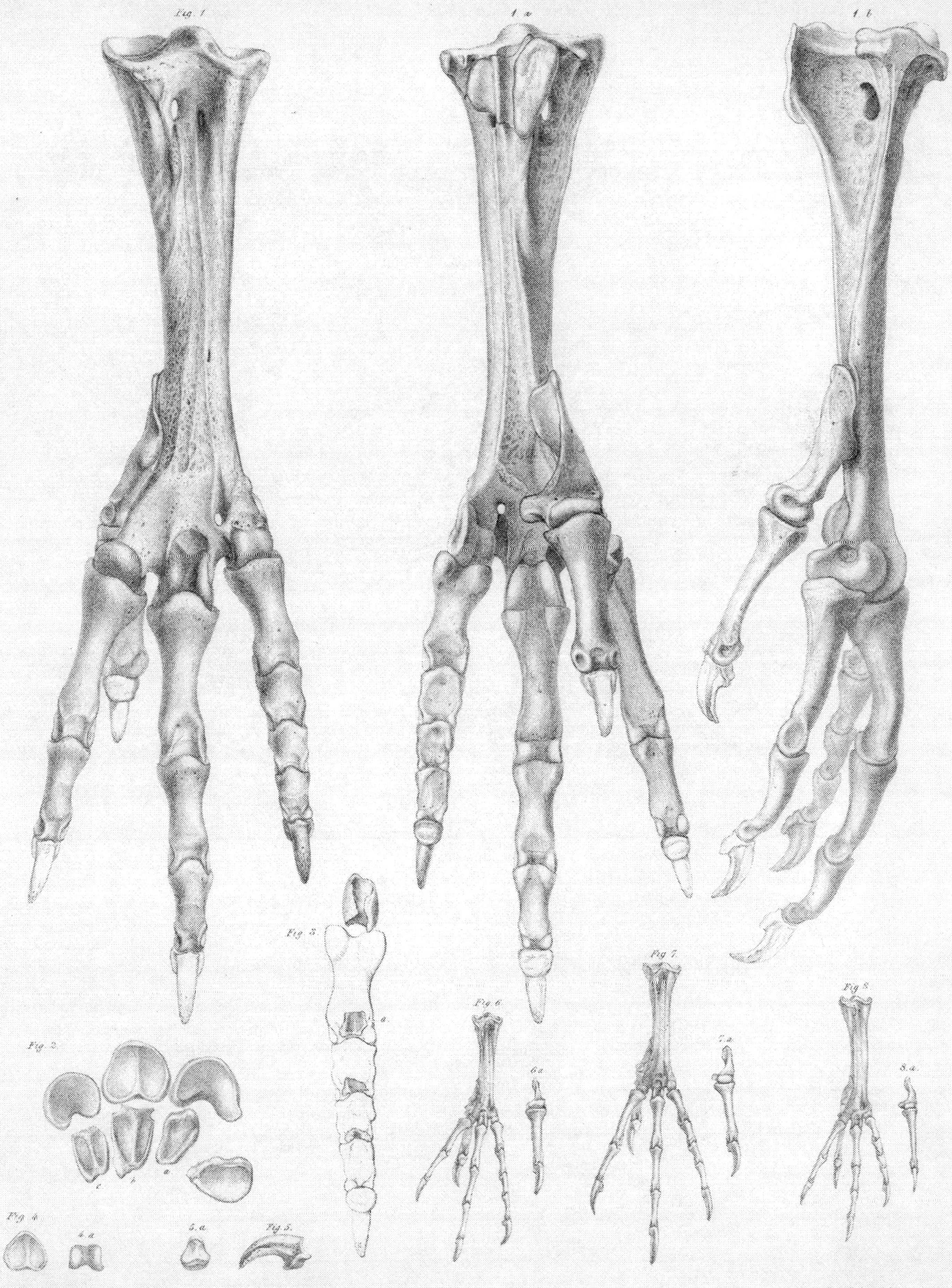

Tuffen West, lith.

Plate XIII.

Fig. 1.

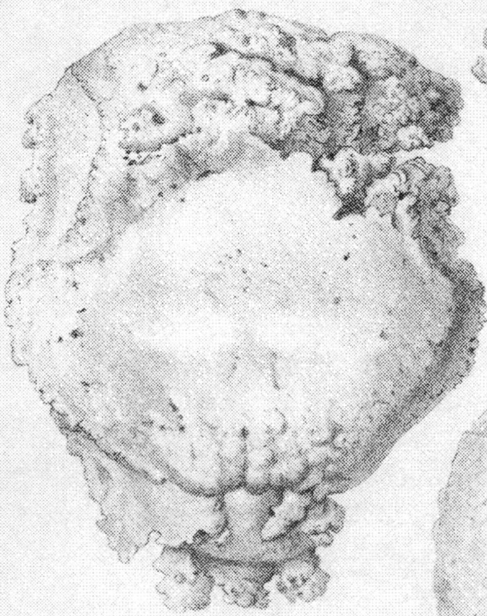

Fig. 4.

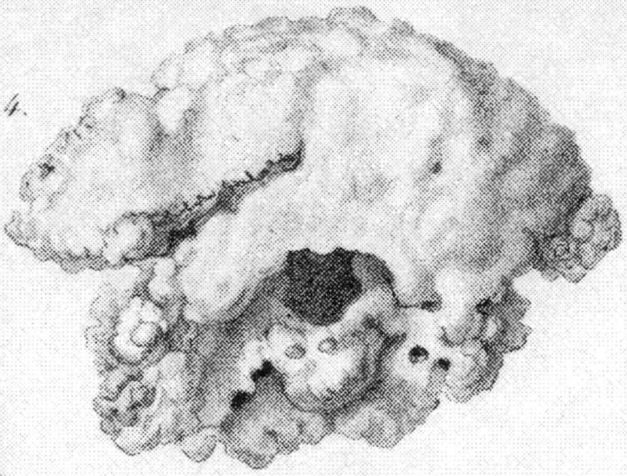

Fig. 2.

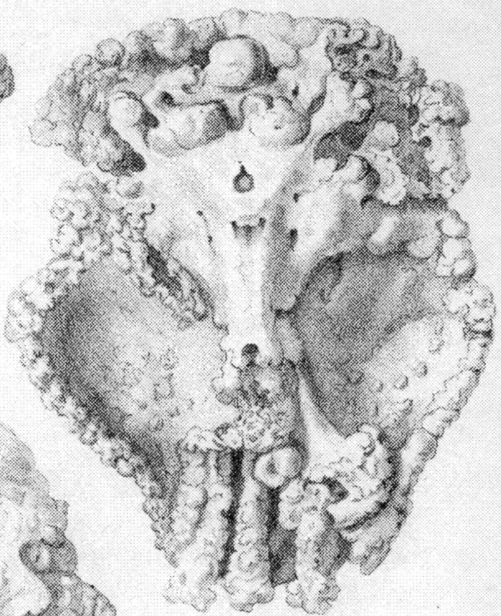

Fig. 3.

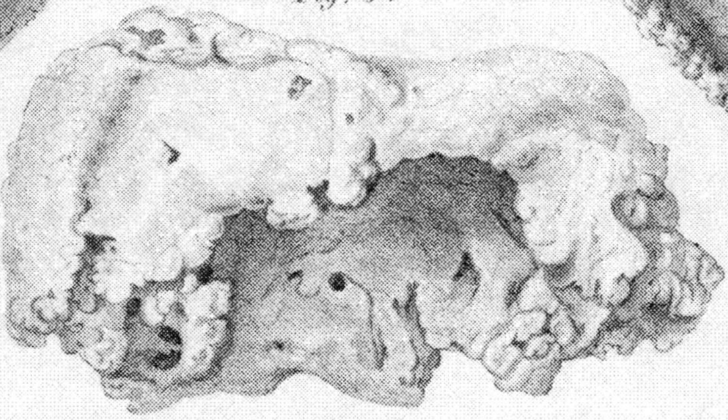

Fig. 5.

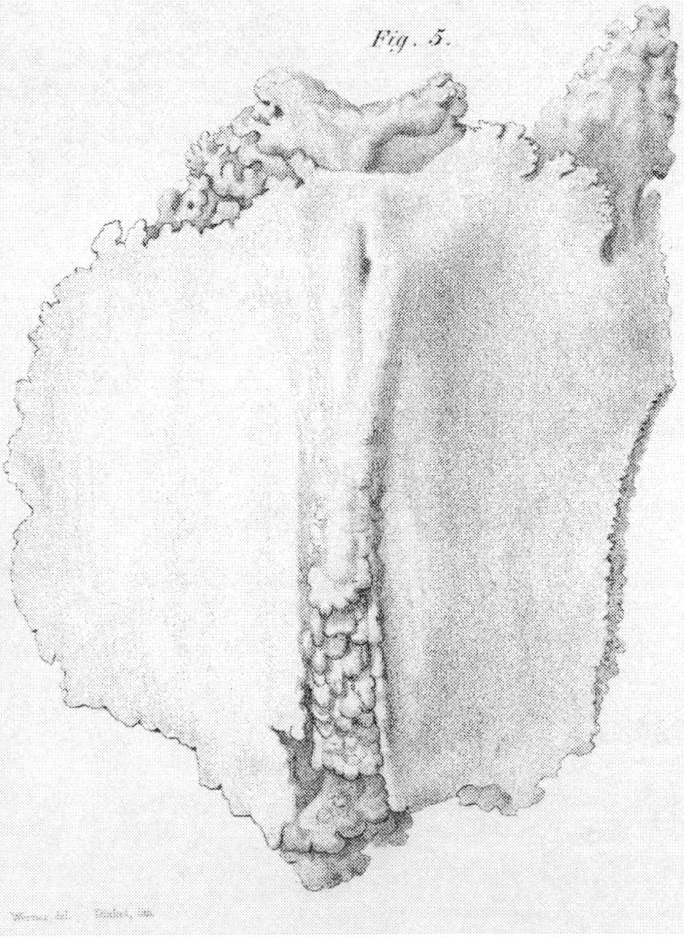

Fig. 6.

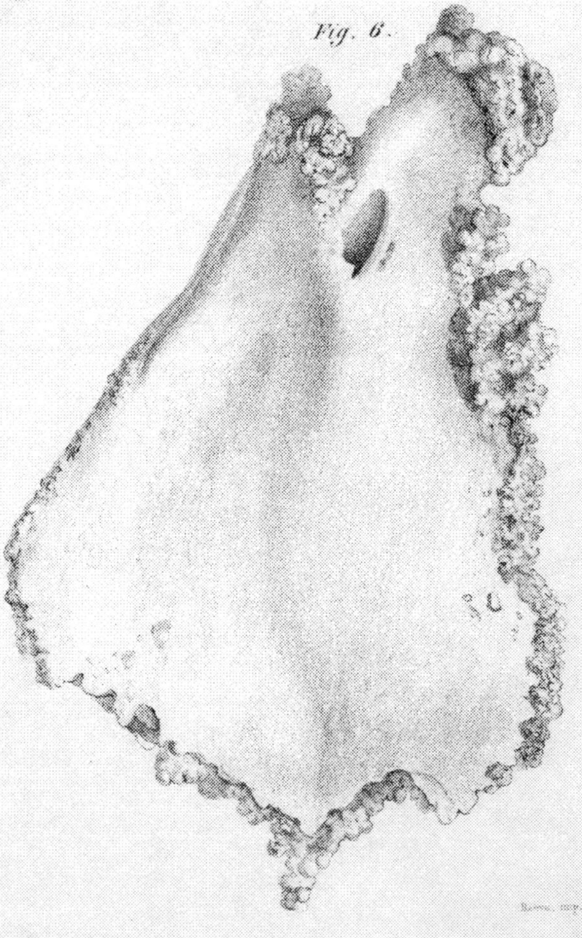

Plate XIV.

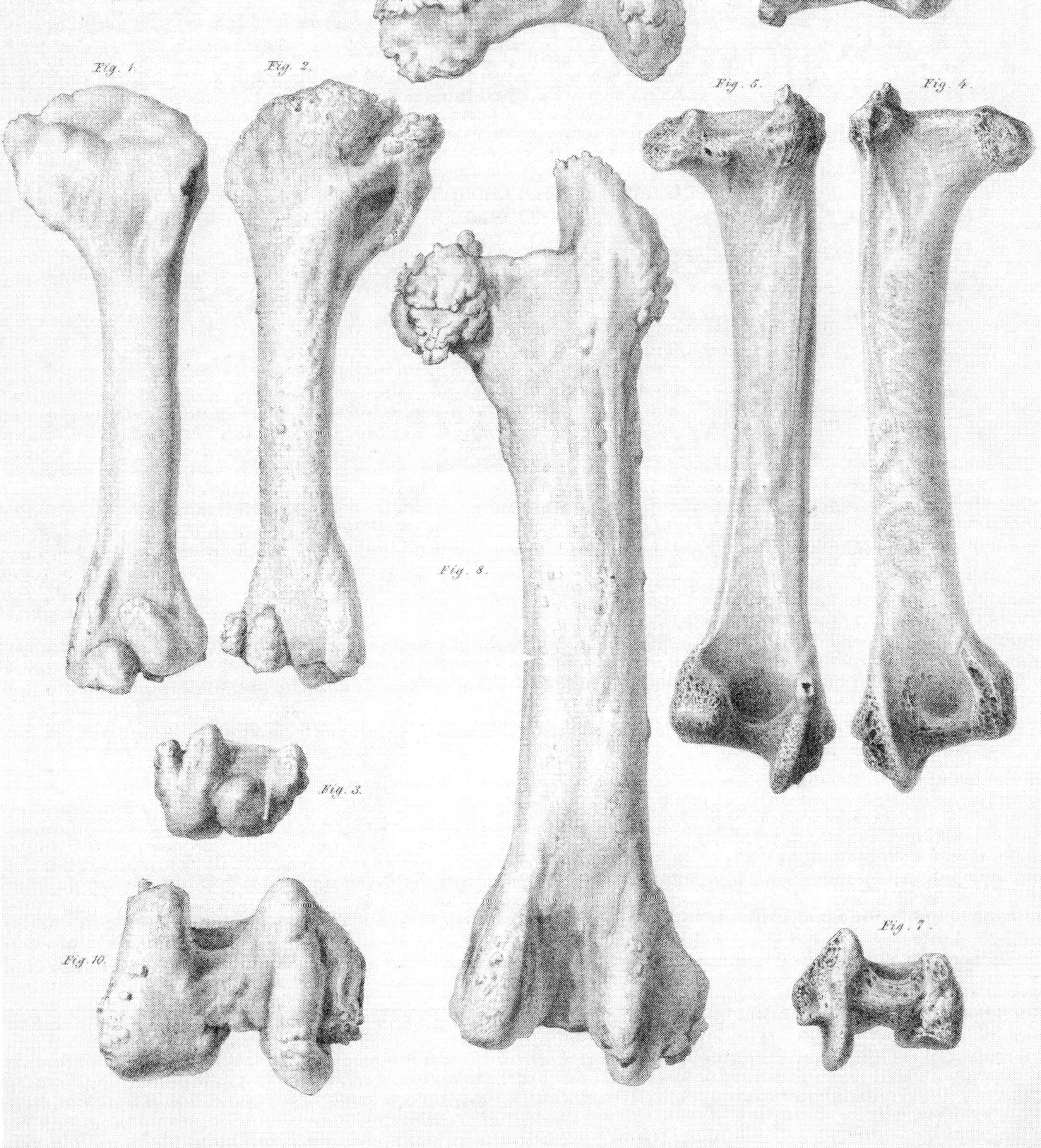

Werner & Dinkel del. Dinkel, lith.

Reeve, Benham & Reeve, imp.

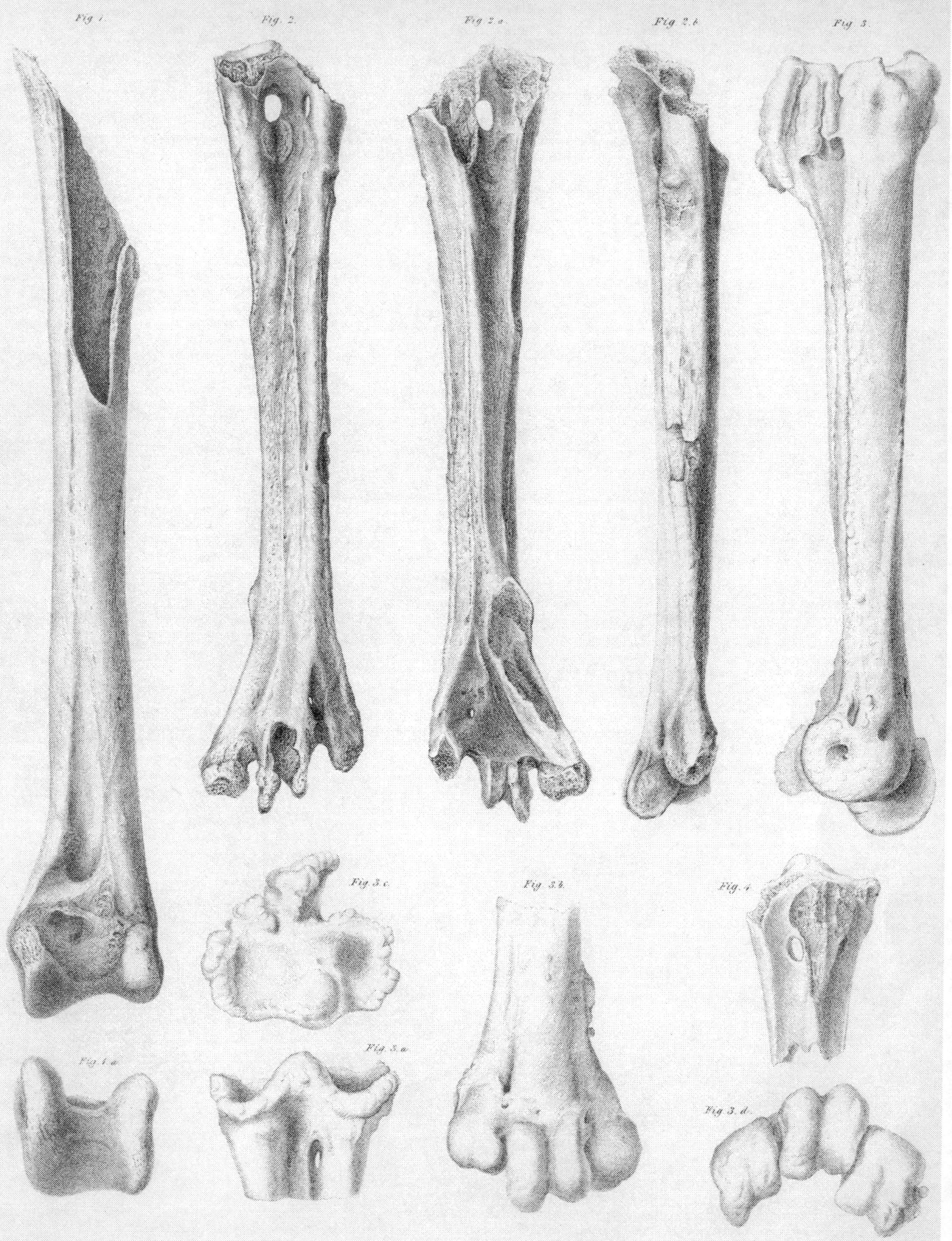
Fig. 1.
Fig. 2.
Fig. 2. a.
Fig. 2. b.
Fig. 3.
Fig. 3. c.
Fig. 3. b.
Fig. 4.
Fig. 1. a.
Fig. 3. a.
Fig. 3. d.

Printed in the United States
By Bookmasters